NO ORDINARY MAN

NO ORDINARY MAN
GEORGE MERCER DAWSON

Lois Winslow-Spragge

Edited by Bradley Lockner

NATURAL HERITAGE/NATURAL HISTORY INC.

No Ordinary Man
Published by Natural Heritage/Natural History Inc.
P.O. Box 95, Station O
Toronto, Ontario M4A 2M8

Editor: Bradley Lockner
Cover Design: Steve Eby
Interior Design: Derek Chung Tiam Fook
Printed and bound in Canada by Hignell Printing Limited, Winnipeg, Manitoba

Canadian Cataloguing in Publication Data
Winslow-Spragge, Lois.
No ordinary man: George Mercer Dawson 1849-1901

Includes bibliographical references and index.
ISBN 0-920474-61-6

1. Dawson, George M., 1849-1901. 2. Geologists - Canada - Biography. I. Lockner, Bradley John, 1950 – II. Title

QE22.D3W55 1993 551'.092 C91-094843-7

Cover Photographs: (clockwise from upper left)
Sioux Camp, Turtle Mountain, Manitoba, 1873.
Oxen Train leaving the Depot, Eighty-One Miles West of Red River.
Bridge over Souris River, Manitoba, June 1874.
Oxen Train and Horse Train, Fifty-three Miles West of Red River.
Painting of G.M. Dawson.

CONTENTS

FOREWORD

Never before have Canadians been in greater need of inspiration or a guiding light as they are today. Perhaps Georger Mercer Dawson's indomitable spirit, his courage and determination against heavy odds, his love for truth and beauty, his high moral standards will bring strength and conviction to those who read his story.

Under the title *Life and Letters of George Mercer Dawson*, my mother, Lois Winslow-Spragge, in 1962, privately published one hundred copies of Dr. Dawson's story for the benefit of her family and friends. The History Department of the Public Archives of Canada at that time expressed gratitude to my mother for the wisdom she showed in preserving George's papers and in publishing the story of his life.

Now in 1993, through Natural Heritage and publisher Barry Penhale's keen interest, George Mercer Dawson's descendants are pleased that this new expanded edition of my mother's book will be available to the general reading public.

George Mercer Dawson was indeed 'No Ordinary Man.' To read about him is like taking a drink of water from a cool, unpolluted spring. His sense of values was so great that he once said he didn't care much for money or possessions. All he wanted was what he could hold in his canoe.

Anne V. Byers
Ottawa
Great niece of G.M.D.

EDITOR'S INTRODUCTION

As one deeply involved in research on George Mercer Dawson for over a decade, it is indeed a pleasure for me to edit this new edition of Mrs. Lois Winslow-Spragge's work. Her obvious admiration of her beloved uncle made the earlier volume a touching tribute to "the little Doctor." Now, in the present edition it is hoped a wider audience will come to appreciate George Mercer Dawson, one of Canada's most remarkable yet unsung heroes. While George Dawson is often lauded as Canada's foremost scientific mind of the latter nineteenth century, few Canadians know the man except for his namesakes such as Dawson Creek, British Columbia, and Dawson City, Yukon.

Who was George Mercer Dawson? Dawson was the pioneer geologist of western Canada. He trekked through vast tracts of that largely uncharted territory from his appointment with the Geological Survey of Canada in 1875, making precise observations which were later synthesized into the foundational geological studies of the area. He was also distinguished as a leader, and was director of the Geological Survey of Canada from 1895 until his death in 1901. George's geological expertise was matched by accomplishments in a variety of other subjects including flora, meteorology, and especially ethnology. On his explorations, Dawson took a particular interest in Indian cultures and purchased numerous artifacts for inclusion in such museums as the McCord in Montreal. His detailed field observations formed the basis for several well-respected studies of Indian life. In whatever subject he addressed, George brought to bear a uniquely gifted ability for analysis and a penchant for exacting detail. His intellectual attainments are astounding.

Behind George Dawson the scientist, is a remarkable person. Along with his mental capabilities George displayed a confident social presence. Even while a young London student at the Royal School of Mines, George socialized with his eminent professors such as Charles Lyell and T.H. Huxley. Later, on western expeditions, he intermingled with Victoria's social elite, including the Creases and the Helmckens. In spite of this extensive socializing, he freely but confidentially admitted that he disliked social events! As his scientific stature rose, Dawson was also warmly accepted as a colleague by the most respected scientists of his era. He held many positions of influence in prominent scientific organizations and won numerous awards and distinctions. Dawson was also renowned as a witty companion who could keep audiences enthralled with tales of frontier adventure. Clearly, George Dawson's intellectual acumen was complimented by a warm and engaging personality.

George Dawson, however, remained an intensely private individual who developed few close relationships. Dawson's immediate family was always the focus of his life. From the childhood admonitions of his devoutly Christian grandfather James Dawson, to the mature interchange of ideas with his father J.W. Dawson, family members were his profound influences. While close to other family members, George maintained an especially intimate relationship with his sister, Anna.

Their childhood sharing of numerous interests grew into a mutually loving and respectful friendship. They remained the deepest of friends even though she married and had nine children, and he travelled extensively. George was always involved, offering emotional and financial support to his cherished sister. Unmarried and never living in his own home, George found lasting personal sustenance within his family. The Dawson family provided George with a firm base upon which to develop his rare capabilities.

George Dawson lived life to its fullest despite questioning his own abilities and motives, and in spite of obvious physical limitations resulting from a deforming childhood disease. His brilliant intellect vigorously applied, produced a substantial body of excellent writings. His stupendous endurance in field explorations vividly illustrated George's amazing stamina. And, his obvious and affectionate concern for family revealed George as a tender and loving brother and son. Indeed, George Mercer Dawson was an "extraordinary" man.

In the present edition the editor has built upon the earlier volume of Mrs. Winslow-Spragge. As in the earlier work, this volume is not intended to be an analytical biography of George Dawson but a portrait reflecting his interesting and varied life. Significant departure from Mrs. Winslow-Spragge occurs where the present edition uses complete letters and diaries in contrast to her brief and isolated quotations. Also, some material deemed superfluous to Dawson's major intellectual achievements was deleted. Re-organization of the text's order was undertaken to develop a smoother transition of subjects. Brief footnotes have been added, where possible, to clarify the text for the modern reader. Dawson's writings also have been reproduced as accurately as possible so that minor errors of punctuation, capitalization, and spelling are not corrected.

In preparing the present volume the assistance is acknowledged of the Documentary Art & Photography Division, National Archives of Canada, McCord Museum of Canadian History, Notman Photographic Archives, and John Summers of the Marine Museum of Upper Canada. Also, the editor has utilized much material originally written by Douglas Cole of Simon Fraser University in his and the editor's earlier work on Dawson. Finally, sincere thanks must be extended to Phoebe Chartrand and Rob Michel of the McGill University Archives for their expert knowledge of the Dawson family and Dawson manuscripts, and their unfailingly helpful assistance.

The following editorial symbols have been used in the present edition:

Material cancelled in manuscript <...>
Material inserted in manuscript {...}
Illegible reading [...]
Conjectural reading [conjecture]

THE "LITTLE DOCTOR," GEORGE MERCER DAWSON

Even as a little girl I had a great fondness for my uncle. I sensed at an early age that he was not quite the same as other men since he was so small in stature and had a strange hump on his back. He was always very kind and generous and had such a humorous twinkle in his keen blue eyes. We as a family looked forward greatly to his infrequent visits. Now after all these years, I can still see him sitting in my father's[1] library smoking cigarettes while talking and laughing with the other men of the family in an atmosphere of geniality.

Presently, I have been trying to do justice to his memory by reconstructing his life from his personal letters, reports and other papers. It has been an inspiring and worthwhile experience to reflect on the achievements of this seemingly frail man who, though physically handicapped, was able to brave the elements and accomplish so much before his death at the age of fifty-two. Through the years he gave freely of his zeal, energy, and mental faculties and performed work of utmost importance and usefulness in the development of Canada.

This biography has not been written primarily from a scientific point of view, but rather with the idea of preserving the nature of this early Canadian; a man of strong courage who worked for the advancement of his country rather than consider his personal convenience and comfort. I feel that all who take time to read these pages will be fascinated by the undertakings of this memorable man.

Those living today might find some interest in reflecting briefly on the era when no whites lived on this North American continent. When this vast country was free and uncontrolled, rivers plotted their courses undisturbed, and mountains and valleys held their great mineral secrets. When thousands of buffalo roamed the grassy plains, and other animals both big and small climbed on the mountain slopes and lived secluded in the rich forests, and game and fish were in abundance everywhere. Indians at that time were the only human inhabitants and were masters of this paradise. They paddled the lakes and streams fearlessly in bark and wooden canoes, and hunted their food skilfully with bow and arrow.

How lovely this land must have been so spread out and undisturbed – a whole continent entirely unmolested and protected on all sides by vast blue oceans! It is difficult to contemplate such a time, and it is rather sad that this enchanted age had to come to an end. But, it was so, for men from far off lands, in their spirit of adventure and desire for fame, were pressing forward and creating new changes in the world. Thus, in 1492 Columbus came to these peaceful shores, and from that time on this huge continent underwent a new birth, such as no one then could have foreseen. Explorers from many lands sailed to North America's shores seeking wealth and glory for themselves and their homelands. These intrepid spirits braved many dangers to achieve their desires, and eventually gained an extensive knowledge of the coastal shores and navigated some of the larger rivers. Thoughout this period

[1]Bernard James Harrington (1848-1907), who married Anna Dawson in 1876, was born at St. Andrews, Lower Canada, and educated at McGill and Yale. Harrington was appointed lecturer in mining and chemistry at McGill in 1871 and was on staff there for thirty-six years. From 1872 to 1879 he also served with the Geological Survey of Canada.

of several hundred years, the Dutch, Spanish, French and British all vied for control of this vast and rich new world. Their various wishes to dominate caused many disturbances and grave fights. In 1776, finally, these struggles resulted in North America being divided into two great parts: the United States of America and what would later become Canada.

During these many years, the face of the land had been changed as white settlers arrived, and towns and villages sprung up. The Indians were gradually forced back, and it was necessary to buy large tracts of land from them for these immigrants. The first of the early explorers who came in sailing ships had long since passed away. Their places were taken by others who accomplished great feats travelling over the continent into unknown tracts of land. Then, with white settlement advancing, there came the first geologists, mapmakers, missionaries, prospectors, surveyors, and transportation engineers. Few North Americans today, living in comfort and even luxury, realize how much we are indebted to these early men who tirelessly tramped the countryside, blazed trails in the forests, and paddled rivers and lakes, making observations and geological notes as to where minerals were to be found, railways could be run, and crops could be grown. It would be well to pause in our hurried lives and pay tribute to these pioneers who accomplished these well-nigh impossible feats and who, through their efforts, have made possible our present way of life.

My uncle, George Mercer Dawson, was one of these pioneer explorers who loved everything related to nature's secrets. His life of adventure and unceasing work revealed here, display his keen search for knowledge, a search that still continues in the lives of others, and will do so as long as man strives for knowledge.

Lois Winslow-Spragge

A BRIEF BIOGRAPHY

George Mercer Dawson began life in Pictou, Nova Scotia, being born there on 1 August 1849. When still a little fellow, he went to live in Montreal when his father[2] became principal of McGill College in 1855. George was a robust child, but at the age of eleven or twelve became ill from a severe chill contracted by playing in the cold spring water of the McGill College stream. His subsequent illness prevented further growth and left him with a hump on his back.[3] In spite of these permanent handicaps he never complained nor allowed them to stand in his way, but bravely went forward. Prior to his illness, George spent one year at the Montreal High School, where he took a high place in his classes. After this, frail health made it necessary for him to continue his studies with tutors. This system no doubt cut him off from some advantages, but did give him wider opportunities for pursuing subjects in which he was interested such as: painting, photography, book-binding, making lantern slides, experimenting with chemical apparatus, and even cheese making and baking clay articles in an outside oven. He also operated a small hand printing press on which he printed money or tokens which he gave to his brothers for chores and errands. George absorbed knowledge readily which was carefully stored away in his orderly mind for further use. He owed much to his father, J.W. Dawson, who never ceased to promote his son's interest in science, and who always watched over George's poor health and obtained the best available medical advice.

The east wing of the Arts Building of McGill College was the Dawson family home and what is now the McGill University campus was the grounds of the residence. Here was where George began his first explorations. In the 1860s the campus was not the trim and mellow spot of today, but had a rugged country-like look. A stream heavily overhung with alders wound its way down the eastern side, where the Science Buildings now stand. George's adventures on stream and campus foreshadowed in miniature his greater experiences later, on the roaring Yukon and other great western Canadian rivers.

At the age of nineteen, Dawson entered McGill in 1868 as a part-time student, attending lectures in English, Chemistry, and Geology. While there he wrote a poem on Jacques Cartier which was praised by his instructors. He also gave evidence of his keen love of nature and poetic instinct in a poem describing the view from the summit of Mount Royal:

"Far on the western river lay,

[2]Sir John William Dawson (1820-1899) was one of the most prominent figures in nineteenth-century Canadian intellectual and scientific life. After studies at Edinburgh University in the 1840s, Sir William was appointed Nova Scotia's first superintendent of education. Serving for three years, he resigned in 1853 when seeking a position at Edinburgh. Though unsuccessful, Sir William was unexpectedly offered the principalship of McGill in 1855, a position he held for some forty years until retiring in 1893. Under his leadership, that institution emerged as a reputable centre for teaching and research. He was also active in a variety of intellectual pursuits and was often embroiled in controversy because of his unflinching theological conservatism.

[3]George was suffering from Pott's disease, tuberculosis of the spine, a slow-working and painful disease that causes the affected vertebrae to soften and collapse and the spine to twist and curve.

Like molten gold, the dying day.
Far to the east the waters glide
Till lost in twilight's swelling tide;
While all around, on either hand,
Spread the broad, silent, tree-clad land;
And in the distance far and blue
Long swelling mountains close the view."

Subsequently, George decided to attend the prestigious Royal School of Mines in London, England, in their three-year program in geology and mining. George studied there from 1869 to 1872, excelled, and graduated as a distinguished student.

From the time Dawson began his serious geological work with his appointment to the Geological Survey of Canada in 1875, his many explorations in Western Canada brought great credit to him and his country. Consistently, his reports were of the highest order, bearing evidence of his striking powers of observation and deduction. Though thoroughly scientific they always took account of the practical and economic sides of geology, and, accordingly, commanded the attention and confidence of mining capitalists, mine managers and others interested in the development of mineral resources. When in the field, geology was, of course, the principal object of his investigations. But, Dawson's wide knowledge of collateral sciences enabled him not merely to collect natural history specimens in an intelligent and discriminating way and discuss the flora and fauna of different regions, but also to make important observations on the customs and languages of Indians, keep meticulous meteorological records, and determine latitudes and longitudes.

In 1895, George Dawson was appointed director of the Geological Survey of Canada. Later, in connection with his holding this position, it was said: "In one sense he is the discoverer of Canada, for the Geological Survey of which he has been the chief, has done more than all agencies combined to make the potentialities of the Dominion known to the world."

Because of his many excellent contributions to geology and science Dawson received much acclaim. George received the degree of D.Sc. from Princeton in 1877, and that of L.L.D. from Queen's University in 1890, McGill University in 1891, and the University of Toronto in 1899. He was awarded the Bigsby Gold Medal by the Geological Society in 1891 for his services in the cause of geology, and was also elected a Fellow of the Royal Society. Two years later, in 1893, he was elected president of the Royal Society of Canada, and in 1897 president of the geological section of the British Association for the Advancement of Science at their Toronto meeting. In 1897 he also was awarded the Gold Medal of the Royal Geographical Society. In 1896 he was president of the Geological Society of America, and his retiring address at their Albany, New York, meeting, "The Geological Record of the Rocky Mountain Region in Canada," later published as a bulletin of the Geographical Society of America,[4] was prized as a summary of Dawson's latest views on problems connected with the complex geology of the west.

Many other distinctions, which cannot be enumerated here, fell to his lot. It was said: "It falls to few men to have so many high honors and grave responsibilities thrust on them in so short a period; the succession is probably without parallel in Canada's history; yet it is the common judgement that the honors were fully merited, the responsibilities borne in such manner as to add renown to the country and the crown.[5]"

[4]"Geological Record of the Rocky Mountain Region in Canada," *Bulletin of the Geological Society of America* 12 (1901):57-92.

[5]W.J. McGee, "George Mercer Dawson," *American Anthropologist* n.s., 3 (1901), 160.

GEORGE AND HIS GRANDFATHER

George's grandfather, James Dawson,[6] *lived in Pictou, Nova Scotia. There, in his younger years, when sailing ships plied the oceans, he carried on a most profitable shipping business and became a well-to-do man. Like many others, though, James Dawson suffered heavy losses when steamships supplanted sailing vessels, so many that he lost almost everything. About this time, he started a publishing business which soon became a successful enterprise, so that little by little he paid back his debts. This was an arduous task and greatly to his credit.*

From the following letters, you will see that James Dawson had deep religious convictions and a great love for his grandchildren. These letters written in 1857 provide rich and detailed observations, and reveal many interesting and worthwhile thoughts. His religious admonitions were rather severe but, nonetheless, his kindness and firm beliefs must surely have exerted a lasting influence.

James Dawson to George & Anna Dawson,[7] Pictou, Nova Scotia, 10 January 1857.

Dr. George & Anna

Twas on three weeks ago, if you recolect I told you of my voyage up the St. Lawrence as far as the Little Scattered Town of *Three Rivers* – we left it, about nine o clock in the morning and proceeded up stream against Wind and current – the day looked like rainy, and we could not see far around us for the dense clouds that hung like a pall from the skies.

Very soon after we <we> left Three Rivers, we met one of these [...] large Rafts which I had often heard of, but never saw before It was about the size of one of my fields at the Farm, as our Pilot told us that it covered about 4 acres and would contain more than 10,000 Tons of Timber [we] counted upon 19 Temporary Houses – saw upon it Great numbers of Men, Women & children and pigs, Dogs, & Hens – they had long poles erected at the end of some of the houses, with [...] at the Top which they light in the night time to warn vessels going up or down the River to Reef clear – soon after passing this we entered St

[6]James Dawson (1789-1862) arrived in Pictou, Nova Scotia, from Scotland in 1811 and became a prominent businessman. James was a devout Presbyterian with extremely strong convictions, who was not always popular in the community. After retirement he spent his last years in Montreal with J.W. Dawson's family.

[7]Anna Lois (Dawson) Harringon (1851-1917), the oldest of George's sisters, was also his closest friend and confidant. Even after her marriage to Bernard Harrington in 1876, Anna continued to share an intimate and rich relationship with George. They corresponded regularly and George recurringly offered assistance to his sometimes beleagured sister, who had nine children. Anna remained in Montreal for her entire adult life, eventually dying of a lung tumour.

Peters Lake,[8] which is merely an expansion of the River, about 15 miles long – The water is all very shallow except a channel which runs straight through the middle of the Lake having on each side at regular Distances [many] posts about 20 feet high with [...] on the Top that look like [...] – but are in fact Beacons on light houses to guide the mariner through in dark nights, when in the middle of the lake we passed another great Raft – hence the appearance of the [Houses] all around us looked very queer. The country all around is very low and level, so that we could see nothing of the land at all, nothing but the Tops of the Trees here and there which looked as if they were growing in the Water, when we were nearly through the lake we came up with 5 or 6 small steamers, which were employed in Dredging up the mud into large leavies to deepen the channel under the command of a Captain Bell, whose death you may have since then seen announced in the Montreal Papers – our Captain spoke to him as we passed – poor man he did not know how soon he would be in the grave & yet we are all equally uncertain how soon we shall Die. The rest of my Travels must be postponed till next letter.

I expect to have a new [volm.] of the "*Band* of *Hope*"[9] in a few days and will send you one – which you both can read and let WB[10] look at the pictures – Good boy!! I am thankful that his Pa says he remembers me and Loves me. I will send him a pretty [...] [...].

Give my Love to mama[11] and Miss Bell – yours

James Dawson to George Dawson, Pictou, Nova Scotia, 21 February 1857.

Dr George

I received your *long* letter of 3rd *past*, and I am happy to say that it shows unmistakable symptoms of improvement in your writing, composition and Geographical Knowledge – as to spelling, I can only see one word wrong and I am not sure but if you examine 4 pages of my writing, you will find more errors than one – It will give me great pleasure to see the *well written* copy you speak of sending me. I should like also to see some of the best specimens of yours and Annas Drawing.

If there be a Steam Boat conveyance next summer from Quebec to this, perhaps Papa will bring Mama, Anna & you all with him to see me. If he does so, I will be very glad to see you all – If Papa comes that way and brings you with him will you remind him to take his pencil and make a sketch of some of the beautiful scenery on the south side of the River: this would afford you an excellent lesson in sketching and one that you would not soon forget.

Little Mary Harris some times asks me when George will come back – I give her a sweetie and a Kiss, and tell her that you will come some time and show her some pretty Flowers and Birds in the Garden. When you come back here you will see another Little Boy over there they call him George Crowe. He can not walk nor speak yet, but he is

[8]Lac St-Pierre.

[9]Or *The Band of Hope Review and Sunday Scholar's Friend*, which was a temperance periodical for children begun in 1851.

[10]William Bell Dawson (1854-1944), Dawson's younger brother, also became a well-known scientist though overshadowed for many years by George and their father. William graduated from McGill in 1874 with his Bachelor of Arts, obtained a bachelor's degree in applied science the year after, then went to Paris to the prestigious Ecole des Ponts et Chaussées. Following his studies there William went into private engineering practice then joined the Dominion Bridge Company as an engineer in 1882, staying until 1884 when he accepted a position as assistant engineer for the Canadian Pacific Railway. Finally, in 1884, he began what he considered his main professional undertaking: director of the Dominion Survey of Tides and Currents. Until his retirement in 1924 he recorded and mapped tides and currents in the harbours and on major steamship routes of the Canadian coasts. William married Florence Jane Mary Elliott (1864?-1945) and the couple had three sons and a daughter.

[11]Margaret Ann Young (Mercer) Dawson (1830-1913) was the youngest of four daughters born to a prominent Edinburgh family. Over the objections of her parents, Margaret married J.W. Dawson on 19 March 1847, and left for life in British North America. Although retiring by nature, Margaret Dawson fulfilled admirably the difficult tasks of a university principal's wife and mother to five children. Deeply religious, Margaret enthusiastically encouraged Christian values and a Christian faith in all her children.

beginning to know people and to know what they say, and before you can see him he will almost be able to Walk as he is growing quite fast.

Do you think you could get me a little plant of the Butter Nut Tree and bring down with you or send it with Papa if you do not come yourself. I should like much to see what like a Flower it has got.

I am sorry I did not pick up some of the Nuts when I was seeing you.

Tell Anna that I have her little *Hyderanga* quite alive, it has set up a shoot already an inch higher, and there is a young sprout from the foot of it which will in a year or two, do to remove for another plant. I will expect, every time that Papa writes me, that either you or Anna will include a note to me – this will show me that you do not forget G.

James Dawson to George Dawson, Pictou, Nova Scotia, 28 February 1857.

Dr George

I have a letter from you today – without date. I hope baby & you have got quit of your colds. You should take good care not to get your feet wet among the cold snow and water – you have heard the old Proverb I suppose that Waters a fine one Good Servants but bad master.

Every body knows what a comfort it is to have a Fire when bad weather comes, and the wind blows keen & chilly. I alone say you have found it being pleasant of an evening, when you have been out building snow men & snow Houses, to gather round a nice fire with your parents and brothers & sisters. But suppose that one of these evenings, when you were snug and warm in bed, a blazing spark or coal should have flown into the room, and burned away while no body was near, till you were awakened by the crackling noise, and smoke caused by the House being in flames your Papa [...] to you, seizes hold of your arm and rushes out with you in great haste, just in time to save your life, while the roof of the House falls in and all his property is consumed. – "What"! you would say, can this dreadful work all come from that little fire by which I sat last evening," "Is it possible that *that little* Fire could do so much mischief." Yes, it is even so. You remember how, last year, Burnside Hall, together with a number of Papas Books and specimens were destroyed, from some such little spark of fire.

The English cathedral also of your city was very lately consumed by <the> [...] to some such small Beginnings – Now a Text came into my mind when I heard of these things, and I thought that I would write to you about them, so that you could talk about them to Anna & WB, and so that you might all see how much *sin* is like a fire. The Text is "behold how great a <fire> matter a little fire Kindleth."[12]

To help you understand and remember it, let us take only the three words, "*A little fire*".

There were once two boys who were brothers, they had grown up together, but one was better than the other and more obedient in his conduct than his elder brother. It happened one day that they were out in the fields together, and a *little spark of sin* in the form of *jealousy*, came into the elder one's heart. He did not check it, for he did not like his brother to be better or more beloved than he was. The spark burnt on to *Anger*, and just as smoke shows when there is fire within, so his face was dark and clouded, and showed that *anger* was there. But soon it blazed out into *Rage*, and he lifted up his hand, seized a club, and killed his brother. AH! he did not at first mean to do this; no, but neither did he check the Rage {or fire} of anger, so that it became his *master* and destroyed him; for God set a mark upon Cain, and there he stood like some blackened ruins, never again to become the fair temple which it had once been. Is not *sin* then like a fire.

I must tell you of two other persons, they were Husband & wife; at the time they lived Christians were very self-denying, and were ready to give up a great deal of their property, and sometimes their lives even, for the cause of Christ. Now these two persons wished to be thought very good, tho they were not ready to make the real sacrifice; so a spark

[12]James 3:5.

of sin in the form of *deceit* sprung up in their hearts, and they began to say to each other, "what can we do to make people think that we are giving up all for Christ, without actually doing it?" Now they were very rich people, and possessed some land, so they consulted how they might deceive about it, was not this like blowing the fire to make it blaze, and so it did, for at last they agreed together to tell a direct Lie. you, no doubt remember what the Lie was which Ananias and Saphira told, and what followed; how it was no sooner uttered than the hand of God fell on them and destroyed them both. How like a *fire* was this; as rapid in its progress, as fearfull in its consequences? (More of this next week meantime I send my love to you all

James Dawson to George Dawson, Pictou, Nova Scotia, 7 March 1857.

Dr. George

Grandpapa is thinking of coming to stay with you, which I am afraid will be a great loss to you, as you will by that means lose the fine oportunity you now enjoy of corresponding by letter with me – There is no finer way that I know of for cultivating the mind than by writing plenty of letters about familiar things. We must just try to make up the loss some other way.

In my last letter I was telling you and anna about what a Great *Fire* a little *spark* sometimes kindles. I told you of Cain & Abel and of Annanias & Saphira. There is just one other character to whom I should now like to refer you.

The *spark* of sin in his heart, took the form of covetousness, which means an inordinate love of money. It seemed to be always there showing itself in little things, like a Tiny spark. He joined himself to a band of Humble men, who cared so little about money, that they put all that each one posessed into a bag, for General use; and this man begged that he might carry the Bag, and keep it for the rest. He seemed to like to *feel* that he could Grasp money, and this feeling increased so much, that he could not bear to see the money spent on costly ointment to anoint the feet of his heavenly master. Thus, by degrees covetousness burnt out all his feelings of affection.

He Loved money better than any thing else, and when the Temptation came – "deliver up your master to us, and we will give you thirty pieces of silver" – that master from whom he had received nothing but Love and tenderness – that master with whom he had lived so many years, and to whose instruction and mild councils he had daily listened, – instead of shrinking from the Honied proposal he yielded. The fire of covetousness had been burning so long that he could not check it, and so it raged on to his destruction, for you both know the fearful end of Judas. *He went & hanged himself.*

In these three instances I have showed you how much sin resembles a fire; if you will take your Bibles and search them, you will find many more, and if you read the History of your own, or any other country, you will see, in the lives of wicked men you meet with there, how much *Sin* is like a fire. We shall find too, that the men who did such evil deeds, did not become bad all at once, oh, no; if you will go to them and hear their account of themselves, we should find that there was a time when *sin* was but as a spark in their hearts, and that if they had checked it then, they might have been happy men.

Now there is one thing I want to impress upon your minds. You all possess the spark of *sin* in your hearts, you know it is so, and whether you do or not, this you know that you feel there is always a readiness to do wrong within you, oh, then be carefull to check it in the beginning. Look into your hearts and see whether it is most likely to break out into *Anger*, or *Lying* or *selfishness*, and set to work at once (asking for the help of Gods spirit) to put out the *fire* of *sin*, whatever it may be.

With my love to Anna & WB – I am yours

James Dawson to George Dawson, Pictou, Nova Scotia, 22 April 1857.

Dear George,

I have received your Letter inclosed in one from Papa of 3rd. April, and it is, by far the best written letter you ever sent me. I am very Glad to be thus able to testify to your improvement. – I can read your fine, *Large, Round hand* without my specs, and I hope you will continue to give attention to your writing, because if you ever allow yourself to Decend from writing a legible round hand, to write a *hasty sharp scrawl* in place, you will never recover what you have lost – The wise saying of Solomon, "Train up a child in the way he should go, and when he is old he will not depart from {it}," [13] is no less True in Physical than in moral things – in the works of our hands than in the Thoughts of our hearts.

I am most happy to hear that Papa approves of your progress in *Drawing* – you will, I hope, be encouraged by this to earn a further amount of his approbations.

But you say Papa has bought you a pair of snow shoes, I am not sure that he has done right in so doing. When Grandpa was a young man he was one time journeying along, on a Wintry day, on Horseback, towards Halifax, in company with {a} Mr Patterson and some other Pictonians. – There had been the previous night a fall of snow of about 15 inches which ended with a shower of Rain, that froze into a thin ice on the Top of the snow but not strong enough to carry a man. – After leaving Salmon River, where we had put up all night, we observed the Tract of a person walking on *Snow Shoes*, going in the same direction we were; when we had advanced about 3 miles our Horses all at once took fright and pricked up their eyes & ears, and were like to run off with us. – we looked in the same direction with the Horses and saw some thing *weltering* and *plashing* in the snow – Two of us dismounted and gave our Horses to the others to hold and went ahead on foot to see what it was – and what do you think it Turned out to be? Why, a poor unfortunate Pedestrian who had borrowed Mr Archibald's snow shoes to take him onto [...], and not having any experience in their use – he fell and his head & shoulders went down through the soft snow while his feet and snow shoes remained on the surface, and in that position he would most likely have soon finished, had we not come to his rescue.

I shall not forget your request about "[*Dayands*] *Book*"

Give my love to Anna. WB and Miss Bell.

Yours afftly

The above gives a brief insight into the character of James Dawson, who was evidently a man of keen observation with a variety of interests. He also obviously believed in the instruction of children, even if the practical lessons of life had to be taught with rather vivid stories.

It is also interesting to be able to reproduce a few of George's letters, written at age eight, in reply to those of his grandfather.

April 23d 1857

Dear Grandpapa

I thank you very much for the seeds you sent us. Dr. Anderson[14] has been visiting us and when he was down town he bought me an Album in four languages German Latin French and English. Mama has planted sweet peas in the crocus pot and they are springing up nicely we each have one side. With love to you {and} give {my} love to Agnes.

May 16th [1857]

Dear Grandpa

I have got my seeds sown and Papa brought <yes> Miss Bell Me and Anna {to the mountain} and we got a lot of Aders tongues

[13]Proverbs 22:6.

[14]Possibly William James Anderson (1812-1873) a Scottish born and educated physician and journalist who came to Canada in the 1830s and worked first in Nova Scotia then Québec.

trillium sanguinaria & saxifrage.

there are also great improvements going on in the grounds there ar I think 18. men busy: they have planted upwards of 400 trees carried away the surface stones and are going to make a new fence.

Please give my love to Agnes.

June 2 57

Dear Grandpa
it is very hot just now. some of my seeds are coming up I go out every morning to pull up the weeds from their roots; papa's garden is laid out very nicely. we were very glad to hear that you are a little better and I hope you will be able to go up the Mountain

CHILDHOOD ESSAYS

This collection of some eleven essays written by George at the ages of ten and eleven, when originally discovered, was a neat little bundle tied up with pink linen ribbon. Each essay was folded in three, with a cover of original hand drawn designs in black and white appropriate to the subject. The essays proved to be most interesting, and rather remarkable for a child of George's age. Some of the better essays are included below.

Vegetation.

Vegetation is that part of life which does not [fell] or move the nearest approach to animal life is made in the sensitive and and pitcher plant whenever you touch the sensitive plant it all curls up and the pitcher plant <it> has at its extremity a pitcher like cavity which is filled with water every morning and there is a little lid which shuts every night. Trees are the largest <a> form of vegetable life the highest trees are the palms and the most spreading the banyan the smallest form <of> is <the> mould there are many intermediate between the largest and the smallest. The food of all animals is originally derived from vegetable sources.

Vegetables {also} form a principal part of the food of man.

They also form the principal part in his manufactures.

May 17th 1860

Rivers <and how>

Rivers are those large <masses> bodies of water which flow from the land to the sea. The cause of rivers is the drainage of the water of the land, which water is caused from rain which is evaporated from the sea up into the clouds and which again fals to fertilize the ground.

The names of some of the principal rivers are these the Amazon the largest which was so named from companies of armed women on its banks the Mississippi one of the principal tributaries {of it} is the Missouri, the St. Lawrence with its tributary the Otawa, {the} Dneiper, Drvina, Don, Volga, Danube, Indes, Ganges, Lena, Obi, Nile {and the} Niger the beginning of the Nile has not been yet explored.[15] Fish and many other water animals inhabit rivers though not of the same kind that live in salt water: in many tropical rivers Crocadiles and Aligaters , shelfish also inhabit rivers but not of <so> {such} beautiful colours as those which live in the sea though some especially in tropical rivers are very pretty. Salt water fish such as the Salmon and Herring come

[15]Dawson's assertions about the Nile are correct in that, though the British explorer John H. Speke had been the first European to discover Lake Victoria in 1858 and the outlet of what he conjectured to be the White Nile four years later, the Nile's source remained in doubt until 1875-76 when Charles Gordon followed the river to the lake.

up rivers to spawn. There are many waterfals and rapids and some rivers are so obstructed with them that they are not navigable but most rivers of sufficient depth are navigable The kind of steamboats which sail on large rivers are not suited for the boistrus navigation of the sea: many rivers are obstructed with shifting sandbanks which make their navigation very difficult.

27 Feb 1861

The Indian antiquities (of Montreal)

There have been lately discovered by accident in montreal remains of a extinct species of indians the {indians} were first discovered by Cartier 3 hundred years ago and the present antiquities are suppposed to be of about that date. Some of the principle things found I will now mention There were five or six skeletons found the bowl of a pipe two stones for gringing their corn with some very pretty pieces of pottery of very preety patern several boan implements, such as knives, piercers, and things to mark their potery, the remains of a broken stone ax a bone needle, worn smothe by use.

The principle animals we find remains of are the Deer, Dog, Beaver, Martin, several kinds of fish, Bear, & Muskwash.[16]

April 10th. 1861

Europe

Europe is divided into 16 kingdoms, namely Great Britain, Portugal, Spain, France, Holland, Belgium, Denmark, Prussia, Germany, Norway and Sweden, Russia, Austria, Turkey, Grease {Greece}, Italy, Switserland. Great Britain is the chief naval power in the world and is a great manufacturing country Portugal is a great wine making country Spain <a> the same France makes a good deal of wine also and is the next naval power to England and has a larger army Holland is a very flat country and is diked to keep out the sea and is next to Great Britain in commerce Belgium is famed for lace which is manufactured at Brussels Denmark is a very much cut up country Prussias soldiars are very good Germany is famed for <its good> {excellent system of} education Norway and Sweden for its timber Russia is the largest country in Europe Austria has a great many mineral products Turkey is the only Mohamedan {Mahomedan} country in Europe and the only one governed by a sultain Greece is famed for its early civilization Italy is the great seat of the Roman Catholic religion and, Switzerland for the Alps and its generally mountainous character.

Europe is the great and most ancient seat of civilasation and *also* is the the seat of the most powerful nations in the world.

May 8 1860

The Lion

The lion is called the king of beasts from its superior strength and cunning.

The {African} lion is by some naturalists divided into two species the brown and the black the black lion is by far the fiercest another very dangerous kind is those who have tasted human blood, as they take a liking for it which induces <it> {them not} to seek for any other kind of food.

The lion <is of> {has} a very fierce look with <its> {his} dark <main a> {mane} its {his} glaring eyes and his majestic step.

Lions gain their food by combined strength and cunning as they lie in ambush near some path which they know to be frequented by cattle and when they see any they give a great bound or spring at them which knocks them quite over when he devours them at his <leasure> {leisure} if the lion fails in {the length of} his leap and the object of it escapes he tries it over and over again till he attains perfection.

[16]Probably muskrat, *Ondatra zibethicus* (L.).

As an instance that the lion does not attack unless he requires <it for> food (he attacks however when he does not need food for revenge) I may mention this incident: A gentleman in Africa going out to hunt alone on his way home came upon a large flat rock <he saw> purposed to sleep there accordingly lay down: when he awoke he saw a large lion sitting at his feet he {he raised himself up a little and} attempted to reach his gun which was beside his legs upon this the lion uttered savage growls and advanced a few paces he continued in this manner for two days all this time the gentleman was <sufering> {suffering} intense pain on account of the heat reflected from the rock till on the second day the lion seeing some deer in the distance left <the> him and went after them, by this time his feet were so scorched by the sun that he could not walk but crept along on his knees till some of the <servents> {servants} coming to seek for him found him and conveyed him home.

I could say much more about the lion but I find I have {not much} room left so I think I will stop here.

26 April 1860

George Dawson as a Child, date unknown.

Dawson's Parents, Margaret and J.W. Dawson, March 19, 1847 & 1897.

The Dawson Sons (from left): Rankine, George, and William Bell.

Anna Lois Dawson, 1871.

At the Dawson Summer Home, Métis, Québec. (from left): Bernard Harrington, George Dawson, Bernard Harrington (son), J.W. Dawson, Anna Harrington, Ruth Harrington, Clare Harrington, Lois Harrington, Eva Harrington, Margaret Dawson, Conrad Harrington.

A BOYHOOD DIARY

The following entries were taken from one of George's boyhood diaries written in 1861 at age twelve, when the Dawson family lived on the McGill campus in downtown Montreal. The diary for the most part refers to his studies and his governess whom he is pleased to call "Miss." There are also a number of references to his health and the simple pastimes that boys engaged in before the advent of television!

March 1st. 1861

Miss came today and we got on very well with lessons. After lessons we went out and had a battle with the students after dinner we went into the Baynses[17] and had some fine fun we played at Jacks alive and several other games <George gave me>

March 2nd. 1861

This morning after breakfast I went to town and bought 38 marbles and went out a little to play. after dinner I took Rankine[18] out and had a game of marbles with Ohara,[19] this eavening I feel very tired and uncomfortable.

March 3rd. 1861.

Did not go to church this morning it was so sloppy red a story in Sherwoods[20] before dinner: after dinner I took rankine out for a walk wrote some of my *bible index* and red a little. William Ross came up and took tea with us tonight. It is sloppy thawing weather.

[17]William C. Baynes was McGill's secretary, registrar and bursar from 1856 to 1887 and the Baynes family also lived in the East Wing of the Arts Building (now Dawson Hall) on the McGill campus. Obviously, the children of the two families visited back and forth between family apartments in the East Wing.

[18]Rankine Dawson (1863-1913) was George's youngest brother. He graduated from McGill Medical School in 1882 and after spending time as a medical officer for the Canadian Pacific Railway in Manitoba, he left for further training in London, England. For four years, Rankine acted as surgeon on liners of the P & O Company before settling in London. In 1896, he married Gloranna Coats and they had one child, Margaret Rita. Always prone to depression and instability, Rankine uprooted his family and moved back to Montreal but, never achieving permanency there, they returned to London where Gloranna left him. Depressed, estranged from his family, and separated from his Montreal relatives, Rankine died in a London nursing home.

[19]Dan O'Hara was a boyhood friend who later dropped out of McGill and qualified as a notary.

[20]A reference to the writings of children's author Martha Mary Sherwood, whose works, including *Little Henry & His Bearer & The History of the Fairchild Family*, were staples of Victorian children.

March 4th. 1861.

Miss came this forenoon. I read a good deal before dinner. After dinner I went down town and bought several things when I came home my *feet* were *very wet*. A lady and gentleman took tea with us tonight.

March 5th. 1861.

Got on very well with lessons today before dinner I went out to play a little after dinner we were playing marbles after that Rankine came out with us we went down to the garden and swang and did several other things Grandpapa dug the first of the parsnips today.

March 6th. 1861

Miss came today and we got on very well with lessons. It was <very> cold today <We> I snoeshoed a little and *sailed my boats* we then played *tag round the College* Ohara came in today and they played old man. I had a <very> *bad headache* so I did not play but went to bed early.

March 7th. 1861.

Got on very well with my lessons today after lessons I red a little, after dinner we went out but it was so cold we were always running in to warm ourselves <I> After that I went into Baynses and spent a very happy eavening after I came in I learnt my lessons.

March 8th. 1861.

Miss came today and we got on very well with lessons Had only 8 speaking marks this week. This afternoon I went to see the snoeshose races I saw them giving the prises. The greatest prise was a sivler cup after that we <went an> slid on the ice. This eavening Nina came in I went to bed early tonight for I am pretty tired.

March 9th. 1861.

This morning being saturday miss did not come it was dull and rainy this forenoon I ocupied myself with reading and several other things after dinner I went down town with Grandpapa I bought some sweaties and Grandpapa bought me a walking cane this eavining william & Anna were both out made a List of seeds to be bought for garden.

March 10th. 1861

After they were all gone to church this morning I red a story in Sherwoods after dinner I learnt my question after that I went to the Bible Class when I came home I finished the story in Sherwoods I went to bed pretty late this eavining.

March 11th. 1861.

School got on very well this morning read a little before dinner after dinner which was late Ohara william and me went down town to get some seeds for planting in box to raise for garden when I came Home I was asked into baynses but mama would not let me go after tea I learnt my lessons and then red.

March 12th. 1861.

Got on very well with lessons this morning they were shortened on account of the College races after the races we had dinner after dinner I went into Mrs. Baynses where we had some fine fun first we were telling stories and reading and after tea we were playing birds in the bush I came home at nine aclock and had to write my composition. I am Late of going to bed again as it is ten a clock now.

March 13th. 1861.

it was very stormy this morning. Miss came got on very well. <after> before dinner I read a little in [...] after dinner I went snoeshoeing all up and down the field after tea papa planted the seads to raise for spring Anna is going to read to me tonight After I go to bed.

March 14th. 1861.

Got on very well with lessons today I read a little before dinner after dinner I went snoeshoeing all up and down the field ran a rase with Ohara and we both came in equell after tea we went to a lecture on the microscope and saw some very pretty objects while the lecture was going on a bench broke down there were some very disorderly people there.

March 15th. 1861.

This morning miss came and we got on very well with lessons I had only 5 speaking marks this week between lessons & Dinner I went down town and got back in time for dinner we went out to play after dinner mamma had a party of the [...] to tonight and I spent a very pleasant eavening it is past 12 a clock.

March 16th. 1861.

I read a little before ten oclock I then went down town with papa he bought some flounders to get shells out of A new [...] after dinner we went out to play and made a great many snowballs and anna told a story when we went in papa planted the verbena slips after tea william and me played marbles with my new board.

March 17th. 1861.

I went to church this morning but it was very cold and windy after dinner anna went to the bible class mama read to me & william after tea I read to <us> myself and wrote a chapter in my scripture index. I have a bad headache my wallflower bloomed today

March 18th. 1861.

Miss came after lessons I played marbles with Ohara & gained & after dinner I went out but it was very cold so we soon came in I was then asked to Mr Baynses where we had some fine fun I just came <in> {home} in time to see anna & nina finish an act Verbena slips growing

March 19th. 1861.

Miss came this morning and we got on very well with lessons a little after dinner I went out we cuttered and pulled each other on our slays Grandpapa gave me a new <story> book and a music book tonight papa and mama are out papa got a box from labradoor full of shells.

March 20th. 1861.

Miss came this morning and we got on very well with lessons after Lessons we went <at> out cuttering and eating Haws[21] & after dinner we played at Thief and Hide and go seek Ohara came in today papa lent me his blowpipe and I was melting glass with it. I then went over home with Ohara and brought Anna home who was <of> {at} the Baynses

March 21st. 1861.

Miss came this morning and we got on very well with lessons I read a little and after dinner I played a little and then went down town with mama she bought a wooden horse for rankine after tea I was melting glass and after that I went for william who was at the baynses

March 22nd. 1861.

Miss came this morning and we got on very well with lessons I had no speaking marks this week so papa gave me 7 1/2 got a very bad headache.

March 23rd. 1861.

I went down town this morning with grandpapa he got me a magic lantern for 7/8 after dinner I showed it to the children went over to baynses they had a party played acting poast office and other things.

March 24th. 1861.

Read a little in the band of hope which papa got a few numbers of <two> two of my seeds came up after dinner I went to the bible class after tea I read a little more and went to bed pretty early felt pretty tired today.

March 25th. 1861.

Miss came this morning and we got on very well with lessons after dinner some people came up to see about the west wing after that papa and me went down to the Natural History after tea I showed the magic lantern Ohara & Nina came over to see it some seeds came up today.

March 26th. 1861.

Miss came this morning as usual Mama chose two new tunes for me after dinner as it was raining we did not come out Ohara and Nina came in tonight and we had some fine acting down in the dining room after that I showed them my magic lantern and a picture of a skeleton which <I> {papa} made for it

March 27th. 1861.

Miss came this morning as usual though we did <not> not expect her as it was so rainy I had a bad headache all this afternoon mama played a little music to me and bathed my feet in water to try to cure it after tea I picked a few forams[22] out of water for papa

[21]Probably berries from wild hawthorn or haw, *Crataegus* spp.

[22]Or foraminifera, unicellular protozoa important as zone fossils especially in the Tertiary Period, some two and a half to sixty-five million years ago, where they may be locally present in sufficient numbers to be major rock-building constituents.

March 28th. 1861.

Miss came half an ower early this morning after school I went over to the Baynses to ask for Ohara who was not at school after dinner I went down town and spent twenty pence on paper and envelopes. a gentleman took tea with us tonight after tea I showed my skeleton through the magic lantern.

March 29th. 1861.

Miss did not come this morning as it was good friday after breakfast papa and me went down to the normal school and in our return visited the indian remanes I found an old [...] after dinner I went to see the boys play [...] Nina came in tonight had a headache in the morning but it got better in the afternoon

March 30th. 1861.

Of course miss did not come today as it was saturday I was playing tunes to mama this morning when Ohara came in to ask me to their house I went in and we played [tops] a good while after that we went up to the cupelo and caught Oharas pigon which we fed and put up again after dinner we read a little and did several little things when I came in at five little miss Wilmot was in I showed them my magic lantern I had a bad headache all the afternoon

March 31st. 1861.

went to mamas church this morning after dinner {I went out with rankine} [...] red a little to william and I listened after that mama read me a little after tea I read a story in holiday after noons

April 1st. 1861.

Miss came this morning got on very well with lessons after lessons Grandpapa gave me a top which I was playing with I went out and played top with the boys I got a piece taken out of my top which I put in again after tea I went to a meating of the Normal school asocation they had some very nice music.

April 2nd. 1861.

Miss came this morning though we thought she would not as it was snowing so badly after lessons I read some of Homers ilead after dinner I was spinning my top nearly all the afternoon I practised a little after tea I prepared some microscope objects for papa & one for myself

April 3rd. 1861.

Miss came this morning after lessons I spun my top after dinner Nina Ohara and us went to see some jugling tricks and some canary birds trained to do tricks at the mechanichs hall after that Ohara and me got a drawing book for Ohara and I got a top and string for myself I learnt my lessons and piled away a lot of paper boxes for papa after tea tonight got a new pair of slippers

April 4th. 1861.

As usual miss came this morning after lessons we were playing soldier after dinner we played snowballing and sailing boats and tried to make whistle of alder but could not after tea I put a lot of specimens in boxes for papa I was asked over to Baynses but could not go.

April 5th 1861

Miss came this morning after lessons we played soldier after dinner we played tag with the girls and snowball <with> and top with the boys. I practised half an hour in the afternoon mama read to us a little after tea and when I was in bed

April 6th. 1861.

We played soldier & top this foornoon after dinner <we playe> I had to stay in till 3 after that we played top papa came and took me over to the indian remains after tea I arainged some specemens some of the students came tonight

April 7th. 1861.

I read a little to william and took rankine out for a walk this fornoon after dinner I went down with papa to the bible class but prince followed us and I had to go home with him mama read some to us before tea after tea I learnt a hymn for papa went to bed early

April 8th. 1861.

miss Macdonald came this morning after lessons we played soldier after dinner we played soldier hide and go seek and shortly before I went in I played top after tea I learnt my lessons went to bed preety early as I was tired

April 9th. 1861.

Miss came today after lessons I went out a little while after dinner I went down town with papa he went to Frothinghams and Dawsons I met mama at the bible depositary on our way up we called on Mrs Torrance[23] after I came home I went to the stabels and helped Tom to cut hay after tea had some music.

April 10th. 1861.

Miss came this morning after lessons I went out a little while after dinner I went down town with Ohara & william we bought a bow and arrows and some sweaties when we came home we practised with our bow and <an> arrows and played marbles.

April 11th. 1861.

Had a bad headache

April 12th. 1861.

Miss came this morning got on very well with lessons after lessons we went out with our bows and arrows after dinner we made a dam and sailed boats and played top mama bought me 3 arrows today some ladies and gentlemen were in tonight

On 13 April this diary ceases. The "journal" as George calls it, he evidently made himself. The cover was of light cardboard, with a design of small blue flowers and the pages were held together at the back with a fold of orange cloth. It also contained a set of directions on how to make colours to paint slides for the magic lantern.

[23]Probably Jane Torrance (1812-1875) the wife of David Torrance, a prominent Montreal merchant and later president of the Bank of Montreal.

A VOYAGE TO GREAT BRITAIN WITH FATHER

In 1865, Sir William took George and Anna for a trip to Great Britain. The excerpts below, from George's diary written as he turned sixteen, will give some account of his impressions and activities on this his first trip across the Atlantic to the British Isles.

July 8, 1865

Sailed from Quebec this morning about nine. The St David sailed about 1/4 of an hour before us. Saw the falls of Montmorrency. met Magnet steamer going up, below Murry bay[24] passed Cacouna at 1/2 past 7 saw Marquis house and beach. kept up in sight of St David all day gradually gaining passed her at nine. water phosporessant. weather very fine.

July 9th. Sunday.

had service on board this morning. Sailed down the south coast all day. Made the lighthouse near Gaspe this evening and started across for east end of Anticosti Gaspe on sight on one side and low land of Anticosti on the other. Passed several vessels this afternoon. St David out of sight astern. Had a little sail up this evening.

July 10th.

Had some sail up all day. Saw a whale. No land in sight all Day.

July 11th. Tuesday

Made the lighthouse at the entrance of Belisle straits[25] early this morning. We could not see much of either the Labrador or Newfoundland coast on account of the thickness of the weather. Saw a great many icebergs near Belisle island[26] and passed close to some saw one ship

Wedensday July 12th.

Went slow most of last night on account of fog. We are getting rooled and pitched agood deal today

[24]La Malbaie.

[25]Strait of Belle Isle.

[26]Belle Isle.

Thursday July 13

Heavy sea and head wind disagreable weather

Friday July 14th.

Fine day and fair wind

Saturday July 15th.

Unpleasant wet day with very heavy swell.

Sunday July 16th.

A very fine day with fair wind. Mr Stag preached us a sermon this morning.

Monday July 17th.

Strong head wind and rain all day. At about eleaven o'clock this morning a steamer came in sight going the opposite way, the first sail we have seen since leaving the straits of Belisle; on signaling she proved to be the St George from Glasgow for Quebec.

Tuesday July 18th.

Strong head wing, with heavy sea and rain all day.

Wedensday July 19th.

Made the lighthouse on Tory island very early this morning. When I came up on deck we were in sight of the Irish Coast we sat watching it till breakfast. Began to enter Loch Foyle[27] soon after we had a very fine view passing up, we saw Greencastle and the fort near it, the former all covered with ivey. When we came up opposite to the telegraph station a pilot came on board also a boat for the dispatches. We then fired a cannon and soon after a steam tender called the lion came alongside and took off all the pasangers and lugage for Londonderry. Going out we passed quite close to the shore and had a very fine view of the Giants Causey[28] and several other fine bold clifs and capes near it. We passed two rocks this afternoon with very fine lighthous upon them called the Maidens, and also had a distant view of the isle of Man

Thursday July 20th.

We passed the bell boy this morning, not long before breakfast, and when we came up from breakfast we were entering the Mercy,[29] in passing up we had a very good view of both banks, we got to Liverpool about 7 o'clock and drove to the railway station where we left our luggage, we then drove to Mr Crows, and found his office not yet opened, when it was opened we found that he was out of town but papa had notes from him and from Scotland, we then walked down and saw {one of} the docks, we then walked to the public museum[30]

[27]Lough Foyle.

[28]Giant's Causeway.

[29]Mersey River.

[30]Now the Merseyside County Museum and Library.

and spent an hour or two in looking over it, there was some very interesting machinary, two models of Liverpool, as well as Natural History specimens, products &c. We then went to St George's hall which is a very fine building indeed. We then went back to the Railway Station and had dinner in a refreshment room. We waited in the Station and watching the election which was going on till the train started at 3.40 we arrived at London at 9.15 and drove to Dr Bigsby's[31] we had tea and then went to bed.

Friday July 21st.

{We} Had breakfast this morning at 9 then drove to doctors and afterwards Torn road where Anna was stopping saw her and then went to the bank drove round <seing seeing> doing some buisiness went in to a confectioners and had lunch, and then walked home up Oxford street stopping to get a map of London and make some other purchases, lay down and rested till dinner. After dinner papa went to see Dr Davis[32] and I stayed at home had tea at half past nine and went to bed.

Saturday July 22nd.

Waited till nearly 11 for Anna and Mrs Simpson as soon as they came we drove to the houses of Parliment and spent some time going through them; we saw the throne in the house of Lords, <and> the house of commons, and St Stephens hall where there are statues of many noted men we also went down into a cloister under it which is a restoration of an ancient one which existed there before, we saw Westminster Hall and went into a court which opens off it which is the divorce court, and waited a little while to see what was going on. We next went to Westminster Abby and saw all the tombs of the kings and queens, as well as those of many other great people, a guide conducted us through the chapels where most of the tombs and monuments are explaining them as we went along. We then drove to the [blank in diary] where we saw a great many very fine pictures we then came home and had dinner. About 8 o'clock Dr Bigsby took Anna and I to Madame Toussauds where there are a great many wax figures of noted men and wimen we stayed there till nearly 10 o'clock when we came home, had tea and went to bed.

Sunday July 23rd.

Papa and Anna went this morning to hear Mr Spurgeon,[33] and I stayed at home. in the afternoon, Dr. and Mrs Bigsby took Anna and I to a very beautiful garden near Regents Park to which they are subscribers, it is very nicely aranged and taken care of and there are several conservitaries in one of which there is a beautiful specimen of the Victoria Regina. Papa went out to church this evening but Anna and I stayed at home

Monday July 24th.

Drove over to Regents Park College this morning after breakfast and Papa left Anna and I there with Dr Davies till he went to see another gentleman when he came home we walked to the Zooligical Gardens and we walked all round them when we had seen them all we went back to Dr Davies and had dinner there Papa went out to do some buisiness and left us there all the afternoon, after tea he called for us and we went home by an omnibus, got home at 1/2 past 9 had tea and went to bed.

[31]John Jeremiah Bigsby (1792-1881), a medical doctor, was also a well-known geologist who had done work on Upper Canadian geology.

[32]Benjamin Davies, a British geologist, was a friend of George's father.

[33]Charles Haddon Spurgeon (1834-1892) was a prominent Baptist preacher who had opened the Metropolitan Tabernacle, seating some six thousand worshippers in 1861.

[There is a gap of several days in the diary]

Friday July 28th.

Started this morning about eleaven for Scotland had a very pleasant journey arrived at York about nine went to a hotel and went to bed

Saturday July 29th.

After having breakfast this morning we walked to the Old York Minster Cathedral and spent some time looking at it and its crypts and some old relics which are there preserved stayed to hear the morning choral service. We then went to the museum which has large grounds around it in which are a portion of the Roman wall and the ruins of an old abby or hospital in the museum are a lot of interesting Roman and Ancient British antiquities. We started in the train for Edinborough a little after two and had on the way a fine view of Newcastle from the high leval bridge, and also a good sight of Berwick. We arrived in Edinborough about 9 and after having stopped at Mr Bells went to bed.

Sunday July 30

Went to church this morning with papa and lay down and read in the afternoon.

Monday July 31st.

Started almost immediately after breakfast for the station and I waited there while papa called on a gentleman Started about eleaven for Musselborough arrived there and had dinner. Papa returned to Edinborough on his way to London between four and five in the <afternoon afternoon> {evening} we were watching the honey being taken from the bees.

Tuesday August 1st.

Intended to start at 8 this morning for North Berrick but put it off till ten. When we arrived we walked from the station to the house we are to occupy after dinner Charles and I went down on the beach and found two hermit crabs and a good many shells came up home and then went down again for salt water for the crabs. After tea we all went down on the beach for a walk and found some more shells.

Wedensday <July> August 2nd.

After breakfast Fred Alfred and I went down to the wharf to fish but we caught nothing we then came home and had dinner, After dinner I wrote letters to Mama, and Charly Davies, and a short scrawl to William and posted them. I then went down to the beach where the rest were and we took a walk along the sand we found some shells; hermit crabs, and shrimps for our aquarium; and we also got some whilks[34] 2 which were boild and we ate some after tea.

Thursday August <2> 3nd.

Went down to the pier and fished we caught two came home and rested till dinner; After dinner we all walked along the shore to the westward a good peace found some shells and other things. After tea I stayed at home and read

[34]Or whelks, marine spiral-shelled gasteropods, probably common whelks, *Buccinum undatum*, used for food.

Friday August <3> 4rd.

Alfred Frederic and I went down and had a bathe before breakfast after breakfast we went down and fished we caught nothing worth keeping. In the afternoon we took a long walk to the eastward found several small fishes in the pools, and a nine rayed starfish. Charles came out <to> tonight <he> he arrived after tea and he and Alfred went out shooting and Frederic went with them they shot a rabbit and a hare.

Saturday August <4> 5th.

We went down to the <beach> pier to try and get a boat to go out shooting in but could not get one went along the rocks and then along the beech but they saw nothing to shoot. <We> I went down to the shore with Mrs Primrose and Marian and sat and read there after dinner. About four o'clock Alfred and Charles found a herring boat and they went out in it with Frederic and I round the Craig they shot one sea bird.

<Saturday> Sunday August 6th.

We all went to church this morning and some of them went to church in the afternoon but I stayed at home after they came back we went for a walk up the hill.

Monday August 7

Went down and fished but caught nothing worth keeping at the pier but went over to the rocks and {Alfred} caught one large fish. I stayed at home and rested in the afternoon and we went out for a short walk in the evening

Tuesday August 8th.

Went out to bathe this morning with Alfred and Frederic but the tide was so far back that we could not find a good place but we got some nice starfish, we came back late for breakfast. I stayed at home this forenoon, but went out to the rocks in the evening.

Wedensday August 9th.

I Went down to the rocks this forenoon with Alfred and Fred and as it was spring tides we got a long piece out and got a lot of sea urchins, some sea mice[35] and [rarer] fishes and star-fishes. I stayed at home writing home in the afternoon, Alfred and Frederic went out to fish and caught quite a lot. We all went out for a walk down to the pier in the evening we sat there a while and then went home.

Thursday August 10th.

It was raining more or less all day today I stayed in and read, all the morning Alfred <Charles> {Frederic} and I went down to fish about four caught only one fish worth keeping. I got a letter from Paris from papa and Anna today

Friday August 11th.

After breakfast this morning Alfred <Charles> {Frederic} and I went to the sea urchin rock we got some sea urchins, sea mice Starfish &c after

[35]An iridescent sea-worm, *Aphrodite aculeata*.

dinner we were prepairing what we got in the morning. Alfred and Frederic went to fish before tea but I stayed in. I was pasting on some stamps after tea

Saturday August 12th.

I got up this morning before breakfast and had a bathe with Alfred and Frederic. I stayed in and read in the forenoon, and went down to the sands in the afternoon. After tea Alfred, Frederic, Charles and I went out in a boat with two other gentlemen fishing for mackerell it was very rough and Alfred and Frederic were a little sick, we came back soon without catching anything. Aunt Uncle Frederic and I {then} went for a walk to the East Links.

Sunday August 13th.

I went to church this morning but stayed at home in the afternoon. In the eavening we all went for a walk to the Glen and back by the beech

Monday August 14th.

I went down to the pier this forenoon a while and came up and read a while before dinner. After dinner Uncle, Alfred, Frederic, Marion, and I went up the Law, but just as we came to the top and were expecting to enjoy the view it began to rain we got into the shelter of a ruined cottage till the worst of it was over and then went home. After tea we went out fishing in a boat but caught nothing we had however a very pleasant sail.

Tuesday August 15th

We went down this morning to the pier to try and get a boat but thought it would be better to put it off on account of the weather. Aunt Uncle and Louisa Cleghorn came to spend the day about eleaven but we could not go out all day on account of the rainy weather Frederic and I took a walk down to the pier in the evening

Wedensday August 16th.

I stayed at home this morning and wrote to William, but went down to the rocks before dinner to see how Alfred and Frederic were getting on with fishing. I read a while after dinner and then wrote to Mama and posted the letter I then went down to the pier with Alfred. After tea Fred and I went out in a little boat with some other boys to the sea urchin rock and back.

Thursday August 17

I went down to the pier this morning about 10 o'clock and stayed there a while. After dinner Alfred Frederic & the <& two> two Halls went out for a row round the black rock and back We came back and took Marian in for a while and then landed. After tea we went and sat at Miss {Matthiews} for a while and then went down to the pier coming back about dark.

Friday August 18th.

Alfred Frederic and I got up this morning before breakfast and had a bathe and also gathered some shells, I stayed in and read all the forenoon. In the Afternoon we went down to the pier and sat a while and then went out to the Craig and back in a small rowing boat. After tea I went out with aunt and bought a sponge took a short walk and then came home.

Saturday August 19th.

I got letters from home this morning by Charles and after I had read them We went down to the shore, and Alfred Charles Frederic and I went out to the craig in a boat, and then a good piece to the east of it, Alfred and Charles rowing we picked up a dead solan goose[36] and Charles shot a Norry but we did not catch any fish After dinner we went down to try and get the boat again but could not get it, we then came up and Alfred and I were engaged most of the afternoon and evening skinning the solan goose.

Sunday August 20th.

I did not go to church today, but went out for a short walk in the evening on the shore.

Monday August 21st.

We had breakfast early this morning to let Charles off in time for the train. I went down to the beech after breakfast gathering shells and came up and read a while before dinner. I stayed at home all the afternoon but took a walk after tea down to the pier. Dr and Mrs Wilson were here spending the day today. Uncle went <home> back to Musselbourgh this evening.

Tuesday August 22nd.

It was raining most of today and I stayed in the house <nearly> all day reading &c till after tea when I took a walk down to the pier with Alfred, we caught a young sparrow, afterwards I went out and bought a pencil case.

Wedensday August 23rd.

I went down to the black rock this forenoon with Alfred and Frederic but we could not get on it because the tide was not low enough. When I came home a letter from Papa and Anna was waiting for me. After reading it I began writing home, and after dinner finished writing and, sent my letter away. I then went down to the pier and went out for a sail with Alfred Frederic, the Halls, and George Lees. We went away out to the west of the craig and then came back to the Pier, and landed some of our party, and then went out again, Alfred shot two Norries, and we caught a few fish. After tea I stayed in most of the evening, but took a short walk before dark.

Thursday August 24th.

I got up this morning and had a bathe with Alfred, Frederic, George Lees and Mercer. When I got home I was very tired so I lay down in bed while Alfred was skinning his Norrie; But he went away to the Bass with Marian and Mr Duncan before he had finished it. I then got up and began mine I worked till dinner, and after dinner I took a rest on the sofa for a while and then began again and finished at five, I then went over to Miss Matthiews for tea with Aunt and Frederic. After tea Fred, I and the Halls went town to the shore and were sailing

[36] Gannet, *Sula bassana*.

boats till dark; when we came home they had returned from the Bass and Alfred had brought a young solan Goose with him.

Friday August 25th.

I stayed in part of this forenoon but afterwards went down to the beech calling at Miss Matthiews by the way with Aunt; I was sailing boats part of the time with the Halls and Frederic. After dinner I stayed in and read for a while and afterwards went down to the rocks and read with Alfred, and George Lees. <I> In the evening we went out for a short sail in the small boat, Alfred and Charles rowing

Saturday August 26th.

I stayed in the first part of this forenoon but afterwards went down to the shore with Aunt. In the afternoon I went down to the pier, and the rocks, and in the evening went down to the beech again.

Sunday August 27th.

I went to church this morning, and in the latter part of the afternoon went down with Charles, and Fred for a walk about the pier. After tea I took another short walk to the pier.

Monday August 28th.

I got a letter from Anna today. Alfred and I went out with the Lees in a rowing boat to the Craig and round it; Alfred had his gun and took several shots but killed nothing. In the afternoon Alfred Frederic and I went out again in a boat to the Craig, and shot a rabbit. I also wrote to papa this afternoon In the evening Frederick and I called at Miss Mathiews, and then went down to the pier with George Lees.

Tuesday August 29th.

We intended to go to the Bass this forenoon but it was too windy so we spent most of the morning about the pier. In the afternoon I was down at the pier again, and in the evening we went out in a sailing boat for Mackerell but caught none.

Wedenday August 30th.

I spent most of this forenoon down at the pier. In the afternoon we were all down to go the Bass, but it was too windy; I had forgotten that it was the day for writing home and came up just in time to write a short scrawl to Mama. In the evening we went down to the pier and fished, we caught more than usual. Aunt Alfred and I took supper tonight with Miss Mathiew.

Thursday August 31st.

I was packing up my trunk to go away this evening, when papa and Anna came in quite unexpectedly we went for a walk towards the glen and back by the harbour. Almost immediately after dinner we started to go up to the train, taking a sight of some pretty gardens on our way <and> papa thinking it would be best {for me} to go right on to Edinborough with him and Anna. We arrived in Edinborough about five and spent the Evening in doors.

Friday September 1st.

In the morning papa and I and Mr Bell went to the taylors and to get orders to see the regalia. <and> When we got home there was a letter from mama waiting, After taking a little rest Anna and I with Ellen and Margaret drove to the Castle and saw there the regalia &c, we then walked to Holyrood Palace,[37] and saw all that was to be seen there and then drove home We meant to have taken a drive round the Queens drive in the evening but it was too wet so we stayed in amusing ourselves with games &c.

Saturday September 2nd.

In the morning after breakfast, we along with Ellen and Margaret went to the Botanical gardens, and when we had seen the gardens with the palm house, the greenhouses, &c., we drove to the railway station and after waiting in the gardens for a while, near the Scot Monument we started for Dalkeith, we arrived at Aunts in time for dinner, and stayed with her till six; amusing ourselves in the garden with Croquet &c, at six we started for Musselbourgh, and stayed there till nine, when we returned to Edinburgh

Sunday September 3rd.

I did not go out to church this morning though papa and Anna did. Alfred, and Charles Primrose were here for dinner. In the afternoon I went to Church with Christena, and in the evening Anna and I had a walk in the garden.

Monday September 4th.

This morning Papa and I went to the taylors, and then to the College where we met Christena, and Anna we went through the College Library, and the Industrial exhibition , afterwards we went, and I bought a pair of vases and then drove home leaving the rest to complete their shopping. In the afternoon we were packing up our trunks and in the evening we took a walk in the garden.

Tuesday September 5th.

We started for Liverpool this morning at ten, most of the Bells, and Alfred Primrose were down at the station, we had on the whole a pleasant journey, and arrived, at Liverpool, about six, (nearly an hour late) we got a cab and went to Mr Crows office and he got in, and drove down to the ferry, we got into the ferry steamer, and then took another Cab on the other side and drove to Mr Crows house. After tea we had a walk in their garden.

Wedensday <August> September 6th.

Anna and I went out this morning to gather blackberries, with the Children and papa went into town, he came back about one and after a lunch he and Anna started for Birmingham. In the evening I went out for a drive with Mrs Crow and two of the children.

Thursday September 7th.

I stayed in reading this morning and in the afternoon went over to Liverpool with Mrs Crow we went to the Egyptian Museum[38] and then walked down to the wharf again shopping as we went. In the evening I was reading and walking in the garden.

[37]Palace of Holyroodhouse.

[38]Possibly the Merseyside County Museum and Library.

Friday September 8th.

I got a letter from papa this morning I spent all the forenoon in the house and garden, as likewise the afternoon.

Saturday September 9th.

This morning, I had a letter to poast so I had to go away about 3/4 of a mile for stamps for it, when I came back there was a letter from papa, and, anna for me with one enclosed from Canada After dinner I went out for a drive, with Mr, and Mrs Crow, and three of the children we went to Parkgate, on the Dee, and back which is 16 miles, it rained a little as we were coming home; when we got back Anna had arrived back from Birmingham.

Sunday September 10th.

I went to church this morning with the rest, and stayed at home in the afternoon and evening

Monday September 11th.

This forenoon I occupied myself with reading, and walking in the garden. In the afternoon we went out for a drive, we first went to some quarries which are, about a mile and a half from here, and got some heather, we then drove to Birkenhead Park, we took a short walk in it and then drove home.

Tuesday September 12th.

I stayed in the house and garden today all day, reading &c. In the evening papa arrived from Birmingham.

Wedensday September 13th.

We started this morning after breakfast for the town, we walked down to the ferry, and got on board

The diary ends with George's embarkation for Canada.

SUMMERS ON THE LOWER ST. LAWRENCE RIVER

George's father thought very highly of the health-giving properties of the lower St. Lawrence River, saying that the ozone was the best found anywhere. George, now nearing eighteen in the summer of 1867, was sent to Cacouna just east of Rivière-du-Loup, on the south shore, to stay with his mother who had a cottage for the season. From here, George wrote the following letters to his adored sister, Anna, his beloved and lifelong friend.

George Dawson to Anna Dawson, Cacouna, Quebec, 21 June 1867.

My Dear Anna

I must confess that writing (as I am now) with nothing to say is a sort of ceremony more than anything else. About the only observation which I made on the way down was one which I have often made before: namely the very curious way everyone, the moment they get their foot on board claim proprietorship in the steamer and always talk about "Our steamer" and funnier still the way they take credit for all the steamer does well, saying for example, "*We* are gitting up steam" but if anything goes wrong then it is all the "captains" fault. Just as the steamer got to Murry bay it began to rain furiously and all the people who landed there got wet to the skin or even deeper; it fortunately however held up while we were landing.

The house is just about what I expected it to be, the worst of it being that but one pane of each window opens. The first flat of this house is arranged so: [illustration in letter]

1 Drawing room (center with a flower of trumpets) and the small stove)
2 Apology for dining room.
3 Mamas bed room containing a red faced attenuated chest of drawers
4 My bedroom containing all the modern conveniances including an oven &c
5 enormous oven.
6 gallery (intended to come into fashion when crinoline goes *entirely* out
7 Lean-too kitchen decidedly airy
8 Pantry or antichamber to the before mentioned
9 stairs with a cupboard underneath access to the top shelves of which is only to be obtained by climbing up the front like a ladder.
10 Last but not least the *glass door* on which papa founded all his hopes.

We have a very good view and might have had a much better but that "the man" (on the principle no doubt that we may have too much of a good thing) <had the> has put the F'house between us and the best view, and the milk house in the same position relatively to us and the next best.

Now I have got a conundrum for William to puzzle out in what part of the house can you you see into every room in the lower flat at once?

"The man" is just now making "des ameliorations"[39] in front of our house before he began the strata or rocks (Ahem) were all sticking up among a medly of chips and stones, at an angle just as if they wanted to look in at our front windows.

The musquitoes here are abundant. at home when you see anything flying along in the dusk you look twice at it to see if it realy is a musquito, but here you never need to do that; hit at any thing you see flying, even a flock of dust and when you get it down it is sure to be a musquito. Mama says she wants part of the back of this sheet so I must pull up short. [Note added by Margaret Dawson] Please tell papa that I took those moths down here and that there are two moths out now, one of whom has laid some eggs so I want him to send me the American Naturalists. Or at any-rate the one which gives an account of the management of the eggs and young caterpilars by the earliest opportunity. I sent the A. N.'s down with some other books to the library before leaving. You may distribute <a little of> my love, but only to those who will value it as I am saving it all up down here.

I suppose I need not mention that I still remain.

George Dawson to Anna Dawson, Cacouna, Quebec, 24 June 1867.

My Dear Anna

I had just finished Williams letter and was going to begin yours when (astonishing phenomenon!) an old man with a smiling face, a towsy wig, and a peep show walked in, so after being refreshed by the sight of "fifteen dirty pictures of palaces &c for deux sous par tete" I now commence yours.

I hope you have got on well with your examinations I suppose that they will be over by the time this reaches you.

It is awfuly dull down here I do wish that you would come down, tell Nina if she will come down we will be eternally obliged to her. By the weather we are having here I should think you are having it very hot in Montreal.

Meat is pretty scarce here just now but fish is plentiful we have already had Salmon, Shad, herring, sardines, smelts, and tom cods.

Sophia and Rankine went to the little English church yesterday where there was just a bakers dozen including the preacher.

You must write soon and If you are too buisy to write yourself get it done by proxy. Mind and tell me all about how you and Nina got on at your examinations.

I hope you find all the alterations, and improvements (?) which are going on very conducive to your comfort. Here are the names of some books which I think would be nice, ask papa if he will get one or more of them; Izac Walton's Complete angler, Dicken's American notes, Mugby Junction, and perhaps Dicsons' "New America"[40] though I dont think It would be very nice.

I have read the reports of the Y.M.C. association[41] and so far I think it very like the award of prizes at the Paris Exhibition.[42]

Believe me your ever affectionate brother

[39]Improvements.

[40]Izaak Walton, *The Complete Angler*. . . (Boston: Ticknor and Fields, 1866); Charles Dickens, *American Notes for General Circulation* (London: Chapman and Hall, 1863); Charles Dickens and others, *Mugby Junction* (London: Chapman and Hall, 18—); and William Hepworth Dixon, *The New America* (Philadelphia: J.B. Lippincott & Co., 1867).

[41]Young Men's Christian Association.

[42]The Paris Exhibition of 1867 held at the Champ de Mars, covered forty one acres. Attendance topped six million and there were some forty three thousand exhibitors.

George Dawson to Anna Dawson, Cacouna, Quebec, 27 June 1867.

Dear Anna

I Thought that writing home would have been finished by yesterdays letters but mama it seems has some messages to send and has not time to write herself, therefor I again take up my pen (cil).

You may judge of the difference between Montreal and here, when I tell you that the lilacs are just coming into flower with us; you must not suppose that there are any such marks of civilization around this house it was someone in the village that gave Sophia a nice bunch. [illustration in letter] {You see while mama was thinking of her messages I employed myself in sketching this ladies slipper. Rankine got it the other day. The flower is pink.}

We have had very fine weather down here all the time as yet, and I fancy a shade or towo cooler than that at Montreal.

Messages.

Mary was told that five pounds of coffee were needed, see that it is got, and got at the coffee mills in St Gabriel street, at 1s per pound.

Bring down the bottle of cod liver oil. (Not for yourself you know)

I thought that there were ever so many messages but it seems there are only two, I hope you will not be dissapointed.

If you think there is time ask William to get me 2d worth or so, of *Pulverised Charcoal*, at any druggists.

Another inconvenience here is the want of any gum or other sticky substance, perhaps you might bring some down. A man came here this afternoon with a very large fish, which he was selling in pieces from door, to door; it must have been nearly six feet long when it was perfect, he (the man) said he caught it down near Isle Verte yesterday, he also said that it was good to eat, so we bought a piece, Dear knows how it will turn out. Mama said she thought it was a Sturgeon.

Rankine and Eva[43] send their love to all at home.

Hoping to see you all down here soon, believe me your affectionate brother

George Dawson to Anna Dawson, Cacouna, Quebec, 28 June 1867.

Dear Anna

"Here we are again". We yesterday received your letters, papas letter, and Williams to Rankine all dated the 25.

We were very glad to hear of all the prizes which you have got, you must bring them down and show them to us. Also please congratulate Nina on Those which she has got.

This letter is writen at mamas instigation, principally to inform Papa that the ground, (or rather rocks) around this house is (or are) altogether too rocky for anything approaching to a garden, Mama has however got a stone wall built up with a cart load of earth inside, in front of the house, she wants to know if she could get two or more verbenas, or some other such flower, in pots to put in it. (Mind you in my own private imagination I consider it impossible, not only because of the lateness of the season but also considering the short time you will have to obtain them in when this reaches you).

We are suffering for want of you here, a fine large salmon was offered at the door today and as he (le garçon) would not cut it we could not get it, there not being enough of us to eat it.

You said papa had sent the American *Naturalist*, down by Miss Lamm. I suppose I will get it tomorrow; If he has only sent one, as your letter would lead me to suppose please ask him, when *he* comes to bring *all* the rest relating to the silkworms.

Mama asks me also to tell you that if you have the oportunity she would be glad if you would go to see Lena before you leave.

Mama wants to appologise for having sent the messages so straylingly.

[43]Eva Dawson (born 1864) was George's youngest sister. She attended Montreal High School, graduating in 1877, then married Hope T. Atkin and moved permanently to England. She eventually had three children.

Believe me your ever affectionate brother

Another year passed and George was again down the St. Lawrence, this time at Tadoussac on the east side of Rivière Saguenay as it enters the St. Lawrence, where once more he was staying with his mother.

George Dawson to Anna Dawson, Tadoussac, Quebec, 8 July[44] 1868.

My Dear Anna,

I am now settled at Tadousac at the hotel, there are very few people here as yet but we expect more soon. The people now here are Mrs Ogilvie[45] who lives in St Catherine street I think and her two little children. Her sister from Upper Canada Miss Dow from Montreal, Young Mr Redpath[46] and his wife who used to be Miss Mills, and their child. That queer young Mr Windham his little sister, little brother, and two governesses (one for each of the youngsters).

Mrs Ogilvie is very nice but has rather a fishy expression her sister ditto, with curls and freckles; Miss Dow is rather goodlooking. Mr Windham as usual, little Miss Windham an awful tomboy, and her little brother one of the crossest, noisiest most spoilt youngsters you ever saw.

I had a nice sail in a birch-bark canoe the other day all about the bay.

I do not finish the account of the Unions running ashore for Mama told me before she left that she had asked *papa* to send her letter out to you.

Maria intended leaving on Wednesday but the Union being disabled there was no boat that day so {Rankine and} she left by the Magnet last Sunday evening.

I cannot write you a long letter for I have to write so many this post, I shall not write any more to you at the Browns for I suppose you will be going home soon.

Please remember me to *all* the Browns and thank Dora for her correction with regard to the "Lightning Bugs"

Believe me your affectionate brother

George Dawson to Anna Dawson, Tadoussac, Quebec, 22 July 1868.

My Dear Anna,

I got your letter last night saying that you intended to return to Montreal on Thursday, & this one I hope will reach you soon after your arrival. You spoke though rather indefinitely of coming down here, If you do I would not advise you to come at present as the hotel is now quite full, & more people coming down tonight, & what they are going to do I am sure I dont know. It would be very nice if you could come down for a little while but under the circumstances I would not advise it at present.

You need not always be writing to ask if I am not lonely, for I am not at all, as yet. I suppose Mama has showed you the programme of the "Concert" I went off very nicely indeed & the "negro melody in costume" was encored. There were about a hundred & twenty people, & 30 dollars 50 cents were taken. We had the room nicely decorated with spruice, flags, &c. Mrs Lemon whose name you will see in the programme, is <a> very nice, from Guelph; and sung beautifully. Mr Holmes who sang the negro song, is an american here on his wedding trip, he is thirty five & his bride only fifteen the cruel parents would not consent, so she went out to go to sunday-school one day & ran away with him. Quite romantic? Both the Lemons & Holmeses have

[44]In the letter the date reads 8 June but at that time George was in Montreal.

[45]A member of the well-known Ogilvie family of Montreal, owners of a large milling and grain merchant firm.

[46]William Redpath, the son of Peter Redpath a prominent Montreal businessman, later died suddenly at a relatively young age. See p. 137.

gone away now, however. The people here are always changing, something like a kalidescope changing a little every day & never coming round exactly the same as at first.

When you write will you please answer a few questions Has Nina come home? Has O'Hara gone away, & if so where? Have the Barnards gone, & if so where; &c. I hope by the time this arrives your great heat will have somewhat abated, and allowed you to enjoy the pleasures of summer without so many of its pains.

With love to all believe me your affectionate

George Dawson to Anna Dawson, Tadoussac, Quebec, 4 August 1868.

Dear Anna,

I got a letter from you last night for which I am much obliged, it was a great deal delayed however by the fog, that is to say the steamer <was> only came in on Monday afternoon instead of Saturday night. You must not expect a very brilliant letter for I was up till twelve last night looking at the stars, and got up this morning at half past three to see a lot of people from the hotel off. We had a beautiful walk over to the wharf just as the day was breaking, and on the way home I went up one of the hills near the Coup sat down there and waited for sunrise, which was beautiful. The only dissagreable thing was the great length of time till breakfast at eight. Such a lot of people are going away now that the hotel is quite comparatively empty. Our second concert came off with great eclat last friday I have not got a programme, but the performers were Miss Rowell, Miss Hickman, Miss McIllree Miss Gough, Miss MacDonald Mr Rogers, and old Mr Hovington who sang two of the queerest old songs you ever heard in the queerest manner possible.

Forty three dollars were collected partly by the sale of small boquets, of which Miss Radford supplied a basketfull. Mr Rogers sang a comic Irish song in character splendidly was encored & gave the Irish jaunting car. We had also a band of musicians from the boat consisting of a harp, flute and violin, who played some very nice pieces. After the concert the chairs were cleared out and they had a dance

Give my love to all at home and believe me your affectionate

George Dawson to Anna Dawson, Tadoussac, Quebec, 12 August 1868.

Dear Anna,

I have received both the paper, and the pamphlets for Hovington safely. I had a very pleasant boat sail a few days ago with a party of people, we went up the Saguenay two or three miles and then went on shore and had "bit of a peruse" getting some very beautiful moss, and returning to the hotel rather late for tea.

I had another very nice sail or rather row the other day. I went out in a small-boat to Lark shoal to look for sea-weeds &c. I had two little french boys to row, and spent a very pleasant time, though without success in the sea-weed line.

Some gentlemen who were up the Saguenay fishing killed a wildcat on their way down, a few miles from here, it was a very large one, and looked as ferocious as could be. I was at the Urquhart's for dinner yesterday and had a very pleasant time. Miss Darey, two of her sisters a few of her children, and a young lady relation, came to the hotel en passant last night, they were on their way up the Saguenay. Some of the Taylors I dont know how many are coming here tonight.

A pic-nicking party are getting off at present for the lake, comprising the greater part of those in the hotel and a good many others

Give my love to all at home, and believe me your ever affectionate

George Dawson to Anna Dawson, Tadoussac, Quebec, 20 August 1868.

Dear Anna,

We have had two days of fog, late boats &c. but this evening it is clearing off splendidly. It is a perfectly lovely evening, <and> the shore on the other side is a beautiful magenta purple, and the sky is full of lovely-coloured clouds. I am sitting out on the gallery watching them, change colour, as I write the whole is changed to deep indigo.

I received papa's letter with money all right, on Tuseday evening. Please tell him so.

I am waiting for an answer to my letter to know about going home &c. I suppose I will get it tomorrow night. There are very few people in the hotel now, or in fact in Tadousac.

We had a very pleasant "reading" the other evening by a certain Mr Palmer he read very well and gave us selections from Dickens &c. The people from the "cottages" were here in full force; After the reading they took to dancing which was kept up till late with great pleasure to all concerned.

I had a very pleasant excursion to Lark reef a few days ago but did not get much, I afterwards went to St Catherines bay[47] to look for chrystals said to be there, I found lots such as they were, but that was not much.

Please excuse both the quality and quantity of this letter? and believe me with best wishes to all, your lazy Brother

George

P.S. Just while I have been writing this letter the fog has come on as thick as ever

George Dawson to Anna Dawson, Tadoussac, Quebec, 23 August 1868.

My Dear Anna,

I dont know whether you will be at home or not to get this note, but write it nevertheless trusting that it will find its way.

I intend to start for home next Sunday afternoon, arriving there if all goes well Tuesday Morning I need not repeat all my reasons, which I have given in Mammas letter.

We have been having very fine weather here lately neither too hot or too cold.

I hope that you enjoyed yourself very much at the Abbots,[48] and had lots of boating, &c. I have been quite troubled by Papa's and Mamma's great anxiety about my coming home alone. Please try and persuade them that it is quite unnecessary.

I will be quite glad to get home again, though I like this place very much, and have spent a very pleasant summer

Please excuse this writing as I have a wretched pen. I suppose you will say that you *never* saw one to suit me yet. But really this one is particularly bad. There are very few people in the hotel now and will be fewer tomorrow.

Give my love to all at home, and with best love for yourself

Believe me your affectionate

[47]Baie Ste Catherine.

[48]Probably Sir John Joseph Caldwell Abbott (1821-1893), later prime minister of Canada, who was then dean of the Faculty of Law at McGill.

A VOYAGE BY SAIL TO GREAT BRITAIN

In 1869, George sailed to Great Britain at age twenty to enter the Royal School of Mines in London. His father prepared the way for him by writing to his scientific friends in London. At this time he was provided with an invalid's chair, which indicates that he was as yet not a strong young man. The trans-Atlantic trip George described as "not very pleasant," as the following diary entries will attest:

Saturday Sept 11th. 1869.

Started from Montreal for Glasgow in ship Lake Erie {930 tons Capt Slater,} from Island wharf at 2.30 PM. after waiting on board from 8.30 AM which had been fixed the night before as the time. Just after leaving wharf in tow of two tugs the ship took the ground opposite the Richaleau wharf and reained there till both tugs were made fast alongside and drew her off. Anchored opposite Sorel at 8.30 PM for the night.

Sunday Sept 12th

Weighed anchor at 4.30, and proceeded down the river. passed Three Rivers at <ten minutes to> 10 o'clock. Passed Batiscan at ten minutes to 12 and the S.S. Prussian at anchor waiting for the tide. She caught up to and repassed us at 2 P.M. Passed Quebec at 6.30 and changed pilots without stopping. Passed S.S. Austrian inward bound 7.40 P.M. Wrote home; and sent the letter by river pilot a Quebec. Dropped anchor, half way down the Island of Orleans;[49] to wait for daylight and tide to go through the Traverse[50] at 8.30 P.M. Port anchor in about 10 Fathoms water 30 fathoms cable out.

Monday Sept 13 1869

Got anchor up and under way at 6 A.M. were delayed till then waiting for the tug which had gone to coal. Light winds and hazy. Passed the Besearch of Yarmouth N.S. in tow of Ranger, and Sunbeam at anchor near the end of the Traverse at 1.40 P.M. Pulled in the hawser <and> cast loose from tug Hero, and set sail at 2 P.M. Abreast of Murray Bay 3.30 P.M. 4.30 P.M. We passed the John Bunion in tow of Hero. Passed Wolfville at anchor 4.40. Abreast of Grosse Isle[51] of Kamouraska light at 6.40 P.M. Opposite Pilgrims light {about 1 1/2 miles off} at 9 o'clock. Have had light S. and S.W. breezes since the tug left us, grad-

[49]Île D'Orléans.

[50]Traverse du Nord.

[51]La Grande Île.

ually freshening, and going round more to the west. It has been a very fine warm day with a beautiful warm sunset over the north shore mountains. Made two sketches one of Goose point[52] above Petit Mal Bay[53] and the other of Grosse Island lighthouse.

A great many white porpoises[54] round the ship all the afternoon and evening. About 8 o'clock counted 32 blowing in 30 seconds. Pilot said in conversation that formerly there was a regular cod and halibut fishery off Green Island[55] though they are not now caught in profitable quantities higher than Father point.[56] Also that there were plenty of lobsters there though none are now seen. That sea cows (Walruses) were caught all up the river the farmers using strips of their thick skins for calache straps,[57] and that Mille Vache shoal was called after these animals Wrote a letter home to send off by pilot at Bic. In case there should not be time to do so before he leaves in the morning.

Tuesday Sept 14th

Pilot left us at Bic 5 A.M. Fine westerly breeze. 10.45 passed Champion of Troon bound up. Fresh breezes W.S.W. Barom. 30.35 sympiesomr[58] 30.195 Thermo 64. 3 P.M. Wind chopped into the Northward, braced up and trimmed the sails[59] 7 P.M. Point de Monts light bearing N.W 1/2 N. distant 18 miles. Wind chopped into NE. braced up and trimmed sails. 8. PM Barom 30.35. sympesom 30.410 Thermom 61.

A lovely sunset this evening over Cap de Monts splendid fiery clouds lying like bais across the west. The south shore a beautiful cool lavender. The sky behind the flaming clouds was a bright soft canary colour. Had the mackerel lines over this afternoon but caught nothing. I find the time passes away very lazily, especially with regard to reading. The soft flashing of the water and noise of the sails seems to exercise a mesmeric efect, and keep one from understanding anything but the most simple books. Took first lesson with quadrant today. 8 P.M. going 6 1/2 knots water phosphorescent where disturbed by ship

September 15 1869

A.M. Wind veering northwards 6 AM Trimmed yards[60] and set lower and <top> mast studinsails.[61] 9 A.M. W end of Anticosty in sight. Light bore SE distant about nine miles. 4 1/2 points variation. Noon wind going westward with clear weather Remarkable hill bearing S. by west about 8 miles off. <2> 2 P.M. abreast N point beacon. Breeze freshening. 6.30 W cliff and beacon bore SW. by W. distant 7 or 8 miles. <3 P> Barometer 30.10 at 3 P.M. symp 30.15 Therm 58 1/2°. 8 P.m. Barom 30.10 symp 30.5 Thermo 58°

[52]Cap aux Oies.

[53]La Petite Malbaie.

[54]More likely beluga or white whales, *Delphinapterus leucas* (Pallas).

[55]Île Verte.

[56]Pointe-au-Père.

[57]Or calage, straps for lowering (of sails, etc.).

[58]Or sympiesometer, a barometer in which atmospheric pressure is measured by the compression of a small quantity of gas behind a column of liquid.

[59]The sails had to be "braced" or "trimmed" (angled) to catch the wind more effectively.

[60]Spars to which are attached the square sails, which run at right angles to the longitudinal axis of a sailing ship.

[61]Or studding sails, additional sails set beyond the leech of a square sail in light winds.

Only passed one ship and two schooners all day. We took the Canadian channel[62] N of Anticosti intending to run through the straits of Belle Isle, the wind being favourable. It has been very cold all day. Passed down mid channel to far off to see anything but the outline of the Labrador and Antacosti coasts. Ship roaling and pitching a good deal and some water coming in by the scuppers. Saw some of the splendid cliffs on the N coast of Anticosti through the glass. Had the mackerell lines over in the morning but caught nothing. I tried <on> the towing net, but the ship was going too fast. 8 P.M. wind began to go ['rount] towards N.E. 10 P.M. Wind about N.N.E. ship going 12 1/2 knots 10.30 wind went round a few points more yards had to be braced up.

Sept 16th. 1869

Fresh breezes and clear weather. 10 A.M. Wolf Island[63] bore NE by N. distant about 10 miles. Noon, strong breezes 3 P.M. Barom rising & a head sea coming. Breezes being contrary we gave into Belle Isle straits, and bore away for Cape Ray Latitude at Noon 49° 55' Long 59° 17' All stunsails out. 8 P.M. Barom 30.475 Therm 58° symp 30.275. Fresh breezes and clear weather. S.S. Very cold.

Friday 17th. Sept

6 A.M. Hauled up for cape Ray in all stunsails braced up. Set fore and afters.[64] Several ships under the lee bow coming up the Gulf. Barom 30.5 Therm 56° symp 30.4. Noon fresh breezes from the East. Barom 30.5 Thermom 55 1/2° symp 30.9. All sail set. P.M. Light winds and variable. 8 P.M. Cape Ray bore NNE distant 8 miles. Water very phosphorescent where disturbed by the ship. Had over the towing net and got several rich hauls. Entomostricans[65] Jelly-fish &c. [illustration in diary] S.S. most of day.

12. P.M. Light winds and variable

Saturday 18 Sept

A.M. Light and variable winds. 5 A.M. wind chopped up from N.W. Trimmed sails. 8 Wind veering NE and E Barom 30.530 symp 30.4 Therm 56°. Noon light winds and clear weather. P.M. Employed washing down. 4 P.M. Wind veering SW [ous] starboard stunsails. Lat at noon 47° N Long by chronometer 58° 54' W. Saw [severas] ships schooners &c. A number of small birds, some sparrows, but mostly a small olive and grey flycatcher were about the ship all day; also a hawk which, caught and ate one of them. Caught several of the flycatchers and put them in the cabin where they flew about catching flies for some time. Tried the tow-net this A.M, and had a good haul of very small red bodies with protruding spicules. Watched the sun go down over the edge of the ocean, not a cloud intervening at the moment of contact it appeared thus, [illustration in diary] and also fore some moments afterwards. A {bright} projection seeming to rise out of the water to meet it. After the upper limb had dissapeared for some moments a small conical light remained visable pointing upwards. When the sun was just on the horison the strata of thin mist in the atmosphere made it to assume exactly the {banded} appearance of Jupiter, through a telescope.

[62]Détroit de Jacques-Cartier.

[63]Île des Loups.

[64]The other kind of sails a vessel can have, opposite to square sails, which when set would be in line with the ship's longitudinal axis.

[65]A division of crustacea, small in size with the body segments usually distinct, and gills attached to the feet or organs of the mouth.

Sunday 19th Sept

A.M. Wind veering SW, Trimmed in port stunsails set all fore and afters. 8.30 passed a light ship bound westward. Saw several large whales blowing. 10 A.M. passed a schooner at anchor on St Peters Bank. Noon Breeze freshening. Latitude obs 46° 24' Long Chronom 56° 58' W. 6 P.M. ship sailing 9 1/2 knots sea getting up. 6.30 passed a schooner running to westward. 7 P.M. commenced raining weather beomming thick. Barometer slowly falling since morning. At 7 PM Barom 30.25 symp 30.25 Therm 60°

Monday Sept 20th. 1869

Breeze still freshening with hazy weather and passing showers. 5.30 A.M. Passed Cape Race bearing NNE distant 5 or 6 miles. in main top gal studinsails.[66] Noon, strong breezes and hazy, with a rising sea Barom 29.9 Therm 59 1/2° symp 29.9 Lat by acc 46.53 Long 51.8 W.

6 PM Breeze <freshening> falling and weather becomming thicker. Had cabin fire lit for first time. Blowing fog horn. 9 P.M. sounded and got bottom at about 45 Fath 11 P.M. Breeze almost gone. Tried tow net, and got a lot of the usual luminous entomostricans some jelly fish of this shape [illustration in diary] and some curious transparent animals of this form [illustration in diary]

Tuesday Sept 21

A.M. threatening to blow by sudden puffs and hazy in all fore and afters and light sails.[67] 3 A.M. blowing hard, close reefed the three topsails reefed and furled cources and furled topsails.[68] 8 AM strong gale wore ship Bar 29 symp 29.7 Noon strong gale with a very heavy topping sea. Ship making very bad weather of it. Wore ship to SE. Cabin filled with water first from sail locker, afterwards on the port side large quantities of water continually on deck rushing from side to side as the ship rolls. <ship> Sometimes shipped several heavy seas in succession and righted with great difficulty. Cargo shifting. Cabin stove broke adrift. Mate was nearly washed overboard.

Wednesday Sept 22nd.

{A.M.} More moderate but a heavy sea an barom rising set lower mizen top sail[69] and reefed fore sail 8 A.M. More moderate, set the three reefed up topsails whole fore-sail and reefed main cource.[70] Noon strong breezes, and passing cloudy weather with showers Lat obs 47° 48' Long by chronometer 50.4 W.

Bar 30 symp 29.75. 7 PM Breeze freshening in up mizen top sails

[66]Travelling up the mast from the deck, the sails on a mast of a typical square-rigged ship would be called: course; lower topsail; upper topsail; top gallant; royal; and occasionally skysail. The "main top gal studinsails," therefore, are those set from the top gallant sail on the main mast as for studding sails.

[67]Probably a reference to studding sails, as in "light weather" sails.

[68]Reefing reduces the area of sails by gathering up portions and securing them with ties called reef points. A "close reef" would provide the greatest reduction, leaving the smallest possible area of sail still set. Furling means taking in the sail completely by rolling it up and securing it to the yard. The "topsails" and "cources" (more correctly courses) are sails described in the above note.

[69]A reference to the lower topsails on the mizzen, or aftermost mast. Setting the sails means unfurling them to catch the wind.

[70]This refers to a particular configuration of the ship's many sails: the three topsails (one on each mast) are set with one reef each; the fore course (the lowest and largest sail on the foremast) unreefed; and the main course with a reef in it.

Thursday 23rd. 1869

Set sail. Daylight more moderate but very heavy head sea.

Set whole top sails courses staysail and main top sail.[71] Tried to catch some gulls and petrals which were flying astern with a hook but the ship was going to fast and sea too heavy.

7 PM Bar 30.15 symp 30.15

Very heavy swell from the NE increasing much. took in most of the sails to keep her from pitching into it so much.

Friday Sept 24th.

Daylight more moderate swell more [easy] made all sail Barom 30.1 symp 30.1 Noon more moderate out all small sails.[72] Lat obs 48° 40' Long 45° 20' PM wind veering aft and swell going down. Out all port. studding sails. A great many gulls and petrals were flying astern all day but refused to be enticed by the bait. 7.30 PM symp 30.1 Barom 30.3.

Saturday Sept 25th.

AM Freshening breezes with heavy rain 3.30 am wind veering right aft, in port studdin sails and mizen royal daylight still increasing with a topping sea. 8 am Bar. 30 symp 30 wind veering NW

Noon Bar 29.95 symp 29.95 strong breezes with a very heavy sea <with> wind running N with heavy squalls 2 PM breeze increasing in top gallant sails 4 PM still increacing with heavy and frequent squalls single reeffed[73] the top sails. 6 P.M in outer jib[74] and reefed the main course 7.30 in upper mizen top sail

Lat by Pole Star 50 25 N.

Saw a land bird on the ship today.

Sunday 26th. Sept 1869

Squally wind veering Nd set upper mizen topsails reefed and whole main course.[75] 4 AM out outer jib and reef of fore topsail. 9 Am reef out mizen topsail and set main top gt sail. Noon more moderate out royals[76] and all small sails. Lat obs 57° 7' Long by chro 36° 46' W. 3.30 wind dying away and going round to W 6.30 PM wind freshening slightly from Wd with drizzling rain.

It has been a very fine day but cold. 9 PM Very dark, and water phosphorescent <both not> in ships wake both generally, and in occasional large bright sparks, ship however going much too fast to use tow-net (9 knots) 10.30 PM breeze still freshening and coming away to Sd of W.

Monday 27th. 1869.

AM Fresh breeze from WNW Noon. strong breezes and hazy weather Bar 30 sum 29.9. 6 PM squally wind drawing Northward in Top Gt stud, lower stud, and Main Royal[77] 8 P.M. Heavy squalls Bar and symp

[71]Again a particular arrangement of sails: topsails unreefed; fore course unreefed, main topsail, and a staysail, which was a fore and aft sail between masts.

[72]Likely the upper and lower topsails only.

[73]Most square sails would have three rows of reef points, making for a progressively greater reduction of sail area, referred to as single, double and triple reefs.

[74]Meaning taking in one of the headsails which are forward of the foremast.

[75]Another arrangement of sails: the upper mizzen topsails were set reefed and the main course was set unreefed.

[76]Set the royals sails on two or three masts.

[77]Took in top gallant studding sails, lower studding sails, and main royal sails.

falling. In Top Main [...] & Top Gallt sails. Reefed and furled fore and mizen top sails, furled main top sail and main course. In jib reefed and furled fore cources. Gale still increasing.

11.30 PM Bar 29.9 symp 29.8

Wind still increasing. A great deal of water on the decks, and the ship making bad weather, the grain in lower hold shifts to one side after she has been for some time on one tack, and makes her take in a great deal of water on the main deck. Saw several land birds on the ship this morning grey and black with short yellow bill and thin black legs; they had a ruddy ring under the neck, and were larger than a common sparrow. Saw a shoal of porpoises passing the ship some of them completely jumping out of the water. They must have been going about 12 knots. The water washing about the deck after dark was full of luminous animals, showing like bright sparks. The spray dashing across the ship driven by the wind same as if it was spirited from a hose

Sept 28th. 1869.

Strong gales with heavy squalls forced to run the ship off before it sea being to heavy for anything else & the ship making such very bad weather, as she would be in the trough of the sea when hawled up to her cource. Taking in huge seas and righting with difficulty.

Daylight, fearful hollow sea and strong gale. A rope had to be stretched along bulwarks of [roof] to hold on by. A great deal of water got into the cabin, and ran from side to side as the ship lurched Noon, a little more moderate. Lat 52° 43 Long 27.3 W Set upper reefed main top-sail and reefed fore sail. Lots of sea gulls near the ship. More porpoises seen today. 6 PM wind freshening again from Nd 8 PM Heavy squalls and strong gales. Wore ship, – fearful work in such weather with so much water on deck, – got her head to the sea and close hauled[78] under two close reefed topsails and main spencer.[79] My trunks had to be brought out of my cabin and secured by rudder trunk for fear of water. Notwithstanding the gale it was a beautiful clear star-light night.

Wednesday 29th. Sept

AM Barometer high and rising but gale still continues. 8 AM do. Noon Lat obs 57°. 51 Long 26° 8 W.

4 PM a little more moderate, wore ship and set mizen lower top sail and fore sail, after [lying] too under two reefed topsails, and drifting to Sd and Ed since 8.30 pm yesterday. 6 PM squally and gale rising.

10 PM Bar 30.25 symp 30.05.Mid strong gales with heavy squalls and a very heavy sea.

30 Sept

Strong NE gales with heavy head sea all night. Ship pitching about tremendously and making hardly any headway. 8 AM Bar 30 symp 30 10 A.M. increasing, reefed fore cource and furled it. Noon strong gales and a very heavy head sea Lat 57° 36 Long by chro 25° 10 W. 8 PM more moderate set fore and main course Bar 29.95 symp 29.85. 10 P.M. tried tow-net met with some difficulty on account of the heavy sea on, but made a pretty good catch, including some curious transparent balls with a dark centre and studded with bristles or spicules [illustration in diary]

Friday Oct 1st. 1869

Daylight. moderate, out all sail Noon smart breezes, and squally with a heavy sea on. Lat obs 51° 41 Long by chronometer 23° 56 W.

[78]To sail close to the wind.

[79]A small fore and aft sail set behind the main mast.

Midnight Bar 29.9 sump 29.85. Water phosphorescent where disturbed by ship, also the waves as they break. Wind light veering and hauling several points N and E.

2nd Oct 1869

AM light winds and variable with slight rain. 4 am Bar 29.9 symp 29.8. Noon Lat obs 52° 34 Long by chro 20° 53. 3 PM Wind veering Nd out port studdinsails. Wind variable. 9 PM In all port studdinsails Midnight light winds and variable saw two outward bound vessels on the Horizon today the first of any kind since leaving Cape race. Sea luminous in ships wake and where the waves break, both generally and in bright sparks. Tried tow-net and found it to be caused by great number of small pulmonigrades[80] <about> from 1/4 to 1/2 an inch across. Made drawing of them.

Some porpoises passed close to the stern this afternoon.

Sunday Oct 3 1869

3 AM Wind veering Sd. hauled up SE by E 1/2 E & trimmed yards and sails.[81] Bar 29.9 symp 29.8 10.30 AM Breeze increasing in fore and mizen royals. Noon. Fresh breezes and cloudy, all sail set Bar 29.93 symp 29.85 Lat obs 53° 46 Long 17° 24 6 PM Freshening in main royal and mizen Top Gt stay sail 8 PM Water phosphorescent same as last night but going to fast to try tow-net which does not work well at a greater speed than three knots, or four with water very <...> calm. Danger of tearing it to pieces at greater speed.

Monday Oct 4th.

Fresh breezes and hazy. 8 AM Breeze freshening in flying [...], and main Top Gallant staysail 9 Am In cross tack. Noon fresh breezes and Hazie weather In mizen top gallant sail Lat by acct 55° 3 Long 11° 46

8 PM Bar 30.1 symp 29.9

1.30 PM passed a steamer outward bound. Midnight. Fresh breezes and Hazie. Water phosph. as last night.

Tuesday Oct 5th. 1869

Smart breezes and Hazie.

Daylight a sailor sent up the rigging sighted the Irish coast, the Bloody Foreland and High land Inside of Tory Island Noon. Tory Island. S 1/2 E, distant about 17 miles. Lat obs 55° 33 Long. [blank] 6 PM. Wind variable and fresh, very unsteady.

Passed Inishtrahull bearing S by W 1/2 W distant about 8 miles. In small sails. 7.30 P.M. Islay light E by S distant about 16 miles. Looking out for a tug all day but could find one. Had mackerell lines out without success. Saw several <ships> vessels very distant near Instrahull. Water phosphorescent <in> small starlike sparks. 10 PM Rhins[82] light bore NE by N, Worked ship to Sd.

Wednesday 6th. Oct 1869

Innistrahull light bore NW by W. Inishowen light bore SW about. Tacked ship to Eastward. 5 AM Tacked ship. 6.30 AM Tacked ship. 7

[80]The mollusc subclass Pulmonata, containing an immense number of species including slugs and snails, only a few of which are marine.

[81]Yards and their attached sails had to be angled to catch the wind most effectively. This job would have necessitated most of the crew hauling on the "braces" which pulled the yards and sails around.

[82]Rinns Point.

passed NWd of Rathlin. 9.40 AM Mull of Cantire[83] bore SE by S. Tacked ship 10 AM Mull of Cantire bore SE by E.

10.15 Bar 30.3 symb 30.15. 11.30 AM Tacked ship to SEd. 2 PM Tacked ship Mull of Cantire bearing SSE 5 miles dist. 6 PM Tacked ship 7.30 PM Tacked ship

Beating up the North Channel all day against a strong tide, and going to leeward till 5 PM. Weather thick and hazy. Can't find a tug. Had tow-net over but broke the bottle in taking it in.

[83]Mull of Kintyre.

Thursday Oct 7th. 1869

12.40 A.M. Made a close shave [...] of Ailsa Craig, passing it by about half a mile. A great many lights of ships and steamers but could not make Pladda Light for some time. 1.40 Tug Stork came alongside and the captain came on board.

George's diary ended as he disembarked at Glasgow, then travelled on to Edinburgh to visit relatives before starting south for school in London.

FIRST YEAR AT THE ROYAL SCHOOL OF MINES, LONDON, 1869-70

The following letters were written to Anna during George's time in Great Britain. Here, at the Royal School of Mines, Jermyn Street, London, he obtained specialized training for his scientific career. He became an Associate of the school and won the highly coveted Murchison and Edward Forbes medals for academic excellence.

George Dawson to Anna Dawson, Edinburgh, Scotland,[84]
10 October 1869.

Dear Anna,

Many thanks for your two kind letters, which I found waiting here on my arrival. I hardly know how to begin my letter, or what to write about, as I cannot go into minute particulars of the voyage &c in so short a time as I have at command. I was only very sea-sick for two days, after which I never passed a meal without taking *something*, the voyage was on the whole not very pleasant from the roughness and coldness of the weather, we had a stove lit in the cabin constantly from the time we left Newfoundland till the morning before we got to Glasgow, that is to say constantly except when it went out by mistake in the night, and nearly froze us all.

Young Rimmer was much more seasick than me, partly I think because he gave up to it so much. I never stayed in bed for it, always managing to get up somehow in the mornings, and if I wanted to lie down using the cabin sofas.

A great many of my apples I am sorry to say spoiled before I could get them eaten, they were so extremely ripe that when they once began they all went off together. The preserved milk was very nice both in tea and coffe, and when dissolved in water, did not make a bad substitute for real milk, with porridge, which we had every morning but one all the way across.

We had plenty fresh beef for about two weeks, and then we fell back on corned beef and chickens, with now and then a tin of preserved meat. The routine was preserved meat, ham and eggs, or something of that sort for breakfast. Corned beef, and chicken for dinner sometimes with tongue; and cold chicken with corned beef or tongue for tea. We had hot potatoes twice a day, and plenty vegetables.

The captain had intended to go through the Straits of Belle Isle, and we went to the north of Antacosti for that purpose, but when we got to the entrance the wind was dead ahead, and so we had to put about and run for Cape Ray.

We had a tremendous gale just on the edge of the Newfoundland banks, the cargo began to shift, and we were quite in a dangerous position for some time. During the height of the gale the cabin stove

[84]George had disembarked and was visiting family in the Scottish city before travelling on to London.

broke adrift and the stovepipe fell down and we had some trouble to fix it. A great deal of water also got into the cabin, and Rimmer who was seasick in his berth, suddenly rushed out into the cabin to say that there was a foot of water running about his room. His trunk was fortunately tin so that his things did not get wet. Although the steward was bailing away all the time we could not keep the cabin dry, and when the ship went on the other tack it all ran across the floor in a horrible way.

I learned very little navigation indeed, and in fact quite wasted my time in every way sometimes feeling a little squeamish, and the constant violent motion of the ship, made anything but very light reading unpalatable. When we had quite recovered from sea-sickness Rimmer and I used to concoct famous suppers and we invented some quite novel dishes. We used to make a raid on the stewards pantry after he had gone to bed, and get biscuits apples &c. We used sometimes to toast the biscuits at the stove, and one very jolly way was to put a lot of cheese on top of one and then toast it on the biscuit into a sort of welsh rabbit. We used also to roast apples on top of the stove very often, and one capital dodge was to cut out the core and fill the cavity with sugar before roasting. We had a good supply of Taffie made twice for us once by the cook and once by the steward.

We used also to be able to get as many nuts and rasins as we liked out of the stewards stores so we ocasionally had a "bloat" on them &c. &c. Margaret and Ellen, and the two Mr Kemps were here last night, also Margarets baby, it is enormously fat, and is supposed to be very pretty. Margaret on the contrary is looking quite thin. Ellen and Mr Daniel Kemp are I think almost as pretty a match as you could see, but Margarets husband looks almost too young for her. I went this evening to the P.B. hall[85] here which is in George St just like the one in Montreal, and heard their great light, Dr Wolson I think is his name speak. He speaks very nicely a good deal like Mr Baynes, though a much younger man. Please excuse me for writing this on a Sunday but I have as much planned for tomorrow as I can accomplish, and have to start for London the day after. I will try and write a little note to Papa tomorrow morning. Please tell every body who you think I should have said good bye to, that I wished to be particularly remembered to them.

With much love,

Believe me, your ever affectionate

George Dawson to Anna Dawson, London, England, 24 October 1869.

Dearest Anna,

I think I wrote to you last time so please excuse anything today, for you see what a quantity I have already written. I was at the South Kensington Museum[86] last night, and as I had not seen it when in London before it was all new to me. It is only about 1/2 a mile from here. It is I believe much improved since you saw it. Raphaels original cartoons[87] are now there, immense pictures and all so well known as engravings, that it is doubly interesting to see the originals. The picture galleries I did not get half through, but could write a quire of paper {even} about what I did see. Turner,[88] Mulready,[89] Landseer,[90]

[85]Plymouth Brethren.

[86]Now the Victoria and Albert Museum.

[87]Born Raffaello Santi (or Sanzio) (1483-1520) in Urbino, Italy, Raphael was renowned as a painter. The "cartoons" Dawson mentions, the so-called tapestry cartoons, were designed by Raphael but executed by a team of assistants whom he trained.

[88]Joseph Mallord William Turner (1775-1851) was a renowned and controversial British artist, largely of landscapes.

[89]William Mulready (1786-1863) was a prominent genre painter and artist, most of whose best works were indeed at the museum.

[90]Edwin Henry Landseer (1802-1873) was a British painter who received many public honours and much popular acclaim.

and all the most celebrated painters copiously represented. The first Sunday I was here, I went to the Belgrave presbyterian church which is near, and today I went to Trinity Anglican church which is nearer. It is low church, & so you see I am following out all your instructions. Christina before I left presented me with a beautiful little testament with an [...] strap which is I think quite a new idea. She is a P.B. you know, and I think a little bigoted, something like the McLimonts. She was especially kind to me when in Edinburgh. I am just next door to Mr Etheridge[91] who papa knows, he as Dr Bigsby said "has no Ice to break" and kindly offered me the use of one of his books.

Yours with love

George Dawson to Anna Dawson, London, England, 7 November 1869.

My Dear Anna,

Many thanks for the necktie which I received in good order, in my last letter from home, I think it is beautiful, and it was a first rate idea to put ribbons in the back, as the other was always becomming too long.

I have to stay at home this morning as I have an invitation from Lady Lyell[92] to dinner at 1 1/2 o'clock & as it is a good distance by the underground, & then about half a mile to walk, I was afraid of tiring myself too much. I will go to church in the afternoon however if I get away in time.

Yesterday I am sorry to say I was an hour late, for my drawing at Jermyn St, but then I had some excuse, as I waited, opposite Buckingham Palace, to see the Queen, go in state to the city to open the Holburn viaduct & Blackfriars bridge. She did not come from the palace, but only passed it on the way from the railway station, where she arrived from Windsor. There was not a *very* great crowd, as nearly every one had gone into the "city". Nor did the "going in state" consist of much; she was not {even} in her *state* carriage. The whole procession consisted of, first a detachment of horse guards then two open carriages with the gentlemen & ladies in waiting, Then an open carriage with the queen Prince Leopold, & princesses Louise & Beatrice, The duke of Wellington as Master of the horse, in another carriage & lastly another detachment of the Horse guards. The queen looked quite blooming, as some one near me said, "how jolly she looks." Prince Leopold looked very handsome & princess Beatrice looked much prettier than I have ever seen her in pictures.

"Truth is stranger than fiction" *vide* Mammas letter with regard to Dr & Mrs [Nottes]. They <looked> seemed quite "spooney".

It is not worth sending you a plan of my bedroom, as it is a very ordinary one. It has two windows looking out on the street. A marble topped wash-stand, bureau, & book shelves. Toilet, & one other table, a bed towel rack, two chairs & a closet for coats &c. I usually sit in the parlour, dining-room &c down stairs where there is a fire, & it is very comfortable as Mrs Guest either talks or remains silent as you seem to wish.

Prof Frankland[93] showed us a beautiful experiment the other day. A large glass vessel which contained a dead cat, covered with a layer of charcoal. It had been so for three years, & we could not help laughing a little when he bent over it & began sniffing to assure us that no smell was perceptable. Most of the students assured themselves still further by having a private sniff after the lecture. I cannot

[91]Probably Robert Etheridge (d. 1903) who was a British geologist and palaeontologist.

[92]Lady Lyell (d. 1873), the former Mary Horner, married the renowned geologist Charles Lyell in 1832 and was a devoted and accomplished wife. She frequently travelled with Sir Charles on geological trips and, because of his poor eyesight, was an invaluable aid.

[93]Sir Edward Frankland (1825-1899) was chair of chemistry at the Royal School of Mines and a renowned and brilliant scientist, who broke new ground in the field of structural chemistry and laid foundations for work in the future.

write more at present as I have to write a few lines to Papa & time is running short. Many thanks for your kind devotion to my interests, I cannot help thinking however that you have mismanaged a little in some ways. Which please excuse. When are my photos coming I only want one or two, but please send them first conveniant opportunity.

I got the Bill of lading of my chair & should not have known what to do with it had not Mrs Fisson (*vide* Mammas letter) found out for me; as I am as ignorant as an infant in such affairs {Bill of lading Marked "Payabil" – but it had already been paid.}

With Much love

Believe me your ever affectionate brother

George M Dawson.

P.S. I have hung your photo over my "one other table" at which I sit when writing &c in my room. But please dont imagine that it needs to be hung up at all. By Williams last letter (and the only one which I have received from him) I think that he is becomming cynical & misanthropic please don't let him relapse in that way, but try to cure him by a little poetry, or music, or a judicious mixture of both.

George Dawson to Anna Dawson, London, England, 14 November 1869.

Dear Anna,

Many thanks for your last long and kind letter, dated Oct 28 which came to hand a day or two ago. I went this afternoon to Westminster abby to hear a funeral sermon by the bishop of London, for Mr Peabody.[94] I was dissapointed however, as when I got there though early the Abby was packed full of people, & the doors closed. There was also a great crowd standing at the doors & in the courtyard. By way of making the best of a bad bargain, I went to St Margarets of Westminster which is quite close to the abby, & there heard a very decent sermon, & afterwords quite an unexpected pleasure, of the dead march which was played on the organ in honour of Mr Peabody.

I saw quite a new trade today, namely a ragged little girl (with a pretty good voice however) going up the middle of the street singing hymns, & eagerly picking up the coppers which some people threw out of the windows to her.

The weather here has been very fine so far, and not cold, though once or twice I have seen the puddles in the streets frozen. The leaves are nearly all off the trees.

I went to Woolwich yesterday (Saturday) afternoon, by boat, to see the London clay.[95] As the boat however took a long time getting there, & the last up one was at four. It so turned out that I had only twenty minutes at the clay [&] I did not do very much. The lights all along the river & on the bridges were very pretty in coming back. The whole way by underground rail, & boat; there & back, only cost 1s & 1d.

I read, according to your wish a bit of the Bible every night, & by stages, have got more than half through Matthew.

Believe me with much love your affectionate brother

George Dawson to Anna Dawson, London, England, 28 November 1869.

Dear Anna,

What I am to write about this week I am sure I do not know, Everything has now happily got into such regular routine that I might almost write the news for next week.

Lecture at 10 four days of the week, & laboritory work afterwards generally till about 3. Laboritory without lecture on the fifth & mechanical drawing on Saturdays.

[94]George Peabody (1795-1869) was a well-known banker and merchant, and a prominent philanthropist.

[95]Dawson later indicates that this was a fossil site.

It takes a tremendous time to go through the course of laboritory work, but I suppose I will do so in time. I am now engaged in <That> finding out substances containing one base. That is to say I get a substance, & then find it out by the dry & wet ways. (Going through a regular routine) & writing the whole method [...] in my laboritory note book. You then take up your book for examination & It having proved correct go on to another. I do several every day, & have now done altogether about 15, there are not many more left to do, so that I hope to be finished with this pretty soon. Next comes either the reactions for the acids, or the analysis of mixtures.

Last Monday I received a bill from the agents of the London steam-ship company, for the freight of my chair & other expenses connected with it. I knew the freight had been paid in Montreal & so was obliged next day to leave the Laboritory early, & go away up into the city by Bus to see about it. It seems that in their bill of lading it was marked "payable in London" but on referring to their accounts they found it had been paid in Montreal. I had however to pay 7s & 6d dock charges, entry, &c. On the way back from this tour I went into St Pauls where I knew service was held about that time, & stayed to hear it. The singing was very good, & the whole very pleasant, except the intoning, which certainly sounded stupid, especially as the intoner had a very weak voice.

I take a good many notes at the chemical lectures. I have already filled one note book & began another. At first my plan was to write the rough notes on one side of the pages, & afterwards copy them with additions alterations &c on the opposite. It made rather a mixed sort of book however, & the pencil also rubbed about by carrying it too & fro. So after filling the first book thus, I now take a rough note book to the lecture & copy my notes in another afterwards. I am afraid though that my notes will be so extensive, as to be of hardly any use at examination.

The holidays at Xmass used only to be a week, but I think this year that we will have a fortnight. The Bells very kindly invited me to spend my holidays with them If they were long enough to make it worthwhile to go so far, I think however that I will be able to spend the time very profitably here, & thus also save the expense of the journey which is considerable. I will be able I think too to get more of a rest here than there.

I hope you had a very pleasant founders festival[96] night before last. Rankine seemed to be quite enthusiastic!

Believe me, with best love your brother

George Dawson to Anna Dawson, London, England, 30 December 1869.

{Please tell Eva, that I am very much obliged to her for the very nicely worked present she sent me. GMD}

Dear Anna,

I have to acknowledge another bundle of letters from home, which has just arrived, many thanks for your regularity in writing & the interesting matter which your letters always contain. I have also to thank you very much for the necktie & cuffs which you sent me, they were quite a surprise on Christmas morning.

I have been having quite a gay time in the holidays. On Friday last, the day before Xmas & the first of the holidays I made an excursion to the tower & monument. I went by underground to Westminster & then by boat to London Bridge. The tower is certainly well worth seeing, there are so many old historical antiqui-

[96]Probaby a reference to Founder's Day at McGill which commemorated the birth of James McGill on 6 October 1744. Traditionally, McGill held a fall convocation about that time though the date referred to by George seems rather late.

ties & localities in it. But then it is one of these places where you are hunted through by a guardian who explains? the objects. I also saw the regalia & the Kohinoor (excuse the spelling) diamond.[97] Some of the inscriptions on the walls of the prison in the tower are very curious. There are some quite ornate carvings, coats of arms &c done by prisoners confined there. I went up to the top of the Monument but had not a very fine view as it was not clear. On Xmas day I only went out for a short walk in the afternoon, & then dressed & went to dinner at Sir C Lyells.[98] It was quite a family party. (About 10 altogether) was very pleasant & not very late.

On Sunday evening I went to a P.B. hall near here of which Chistina gave me the address, but was not much pleased. The congregation was only 52 in number, in a [...] stuffy hall, & presented a typical series of the same old faces always observed in such places. (Talk of the *uniformity* of Roman Catholicism, after observing the *flat uniformity* of P.B. congregations!) The speaker was I suppose appointed to the post because his head was bald in the same place as Mr Darby's[99] & his heart of the same cut. Certainly not for his powers as an orator.

On Monday, I went with Mr Guest to the Crystal palace. It was "Boxing day" & I only went to see the crowd. I had a very good chance of doing so as there was nearly 40,000 people there. On Tuesday I went to the South Kensington Museum in the afternoon & to Dr Carpenters,[100] in the evening, to dinner. He had a small dinner party & then some more people came in the evening & raised the number to about 30. It was very pleasant. I left as early as I could but did not get home till long past twelve, as I came by underground & being the last train, & having to change at Baker St. There were many delays. Yesterday I went to call at Col Lyells & return the Chemistry which his {son}[101] was kind enough to lend me, but which he now requires himself. I afterwards called at the Davies's & they asked me to stay for tea, which I did. They had asked two or three friends (or relatives I think) to tea. Their Daughter is somewhat better, though still very delicate. She was in the room & took tea with us, but went to bed early. I got home by 9 o'clock. I Intend to send by this mail if I have time to do it of a brooch (6d) from the crystal palace for Eva. Please excuse the monotony of this letter & Believe me, with best love to all at home & to yourself in particular.

Your Affectionate brother
{many thanks for the photo}

George Dawson to Anna Dawson, London, England, 8 February 1870.

My Dear Anna,

I have no knews, It seems to me as if I was always writing & about nothing. For as I write on Sundays & then finish up the letter on Thursdays, when I sit down to write again It seems as though it was only yesterday since I wrote last. I did not receive any home letter till very late last week & so have now to acknowledge, <a> letters from you & Mamma. Yours being the first after your return from Quebec.

[97]The famous Indian diamond known as the Koh-i-Noor has been worn by all queens since Queen Victoria and is now in the crown made for the coronation of George VI and Queen Elizabeth in 1937.

[98]Sir Charles Lyell (1797-1875) was undoubtedly the most famous nineteenth century geologist. He established geology as a science, especially through such foundational writings as his *Principles of Geology*. Lyell had travelled in Nova Scotia with J.W. Dawson in 1842, and profoundly influenced the elder Dawson's geological career.

[99]John Nelson Darby (1800-1882) was founder of the Darbyites, or exclusive party among the Plymouth Brethren.

[100]William Benjamin Carpenter (1813-1885) was a renowned University of London naturalist who wrote widely in zoology, botany and physiology.

[101]Colonel H. Lyell was Charles Lyell's brother and his son was Leonard Lyell (1850-1926), later a member of the British parliament.

Yesterday (Saturday) after drawing class was over at 1 o'clock, having nothing particular to do I went to see the pictures in the National gallery Trafalgar Square, & spent about two hours there. There are a great many of Turners, which he presented, & also some of Wilkie's[102] & Collins's,[103] also six of Hogarths,[104] & one or two rooms filled with the productions of old Continental Painters, mostly representations of saints, or passages from Scripture. Some of Turners I like very much but others are much too Turneresque, & not being artistic I fall into the vulgar fault of admiring his worst. One of the very prettiest is "The Fighting Temeraine being towed to her last berth"[105] you can really distinguish where the water ends, & the sky begins, & the sunset colours are beautiful. One, supposed to be a beautiful study, is called "rain, steam, & speed"[106] has a viaduct in the foreground with a red & black blotch upon it which is supposed to represent a locomotive, followed by a train, this you see through a thin veil of mist, & in the background is a mass of inextricable, earth, sky, rain, mist & clouds. Two very beautiful pictures are by Sir Joshua Reynolds, "The Infant Samuel" & the "Age of Innocence" prints of these, you see everywhere.[107]

On my way back I walked across St James's Park, & stopped on the bridge across the ornamental water to see the ducks & swans of which there is a very good collection there. Some of the ducks are beautiful & it is most amusing to watch them, swimming & diving about. The swans are black Australian ones, & though not so large as the ordinary swan, are both graceful & pretty. In the park I noticed several bushes beginning to bud out & tips of green showing, the grass is also quite green indeed has been all winter. & The crocuses are showing above ground. In fact I dont think that you can say there is a winter here at all, only an unnecessarilly protracted autumn & spring.

I was reading the other day, somewhere or other, that children should be fed with all sorts of romances, & fairy stories. I think it is quite right, & very necessary for if they are always, matter of fact when young, & have no exercise for their imaginations; they get to loose that very useful power when they grow older, & soon degenerate into, dull, weary, materialists. Is it not so? —-

Everything goes on just the same with me now, Lecture, Laboritory, reading up & writing up my notes in the evening, all very pleasant & not very hard work, but nothing to write about so you see I am obliged to spin out what I have, & very often, as I suppose you have by this time found out, my letter consists of <of> an account of something I have been doing. or some place I have been on the previous Saturday. And now I will not quite finish, in case I find something to say or have another letter to acknowledge before I send this next Thursday Morning.

{Feb 10th.

I have to finish without having got my letter, the mails seem to have got a habit of being late.

I suppose this will reach you sometime about your birthday. My best wishes for many happy returns

Your affectionate brother}

[102]Sir David Wilkie (1785-1841) was a recognized British painter who had studied at the Royal Academy.

[103]Probably William Collins (1788-1847) who was a very popular landscape and portrait painter.

[104]William Hogarth (1697-1764) was a leading British painter and engraver who excelled as a satirist.

[105]The exact title of the painting, done in 1838, was "The Fighting 'Temeraine' tugged to her last Berth to be broken up."

[106]The painting's full title was: "Rain, Steam and Speed —The Great Western Railway."

[107]As Dawson correctly noted Sir Joshua Reynolds (1723-1792) was one of the most popular artists in Great Britain. Especially known as a portrait painter, his prestige and influence were enhanced by his presidency of the Royal Academy which position he occupied from 1768 until his death.

George Dawson to Anna Dawson, London, England, 20 February 1870.

My dear Anna,

My letter last week did not, on account of the stormy weather arrive till Saturday morning, & I have to thank you as usual.

You talk about Chinoguy[108] & his wonderful doings. I think however that it would have been better had it not been arranged that he should stay in Montreal. For though people will come to hear him from the novelty &c. I think that if he stays permanently they will very likely soon tire of doing so. If he made a visit once a year or so, he would always come with novelty & create a sensation. Besides do you think it right that he should desert that colony of Canadians whom he has lead out into the wilds of Illinois? You see to what I am reduced because of the want of news.

I said in my last to W. that all the frost was over, but immediately after writing it came on again & though not so severe it is freezing still, & skating going on. The laboritory in consequence, has been pretty much deserted for the last few days, only about half the fellows there. One day last week the ball-cock of the cistern was frozen & we had no water all one afternoon, except a rather scanty supply of distilled.

You talk of show-shoe walks, or walk, at which you assisted. What jolly ones we used to have last winter, & how I wish I could join you in one; across the mountain, or round its east end. I only hope that they are not cutting down any more of the trees, or that Mrs Redpath[109] is not behaving, (in that respect) in the shameful manner which she did last winter.

The organic Chemistry lectures are getting more interesting again, & Frankland shows his own interest in the subject by always prolonging them at least quarter past the hour. I have bought two new note-books & so have a fair field to begin upon tomorrow.

I have heard of a chance of getting a Franklands chemistry[110] from a second-years student, which, If I can <suggest> accomplish, will be very satisfactory, I will add a note about it, before closing this on Thursday.

You dont' say anything about Papas cold, but Mamma does. I hope it is quite over long ere this.

It is a good thing that the postage to Canada is reduced, or I should feel exceedingly guilty for sending a letter so full of nothing as this one. But somehow I always have to write when not in the humour. I think I must invest in some of Borwicks, or other good baking powder, by which to prevent my letters from being so "flat," & "sad".

Wednesday Feb 23rd.

I have not yet ascertained about the Frankland, but hope to do so next Saturday. I think I have a good chance of getting it.

Believe me with very best love, to yourself & all at home.

George Dawson to Anna Dawson, London, England, 17 April 1870.

My dear Anna,

Have you ever been at Kew?[111] I dont' remember whether you went there when you were here. At any rate I went there last Friday & enjoyed myself very much. Our Easter holidays consist of {Thursday &}

[108]Charles-Paschal-Télesphore Chiniquy (1809-1899) was a controversial Catholic priest turned Presbyterian minister who settled in the parish of Ste-Anne-de-Kankakee in Illinois, after leaving Lower Canada in 1851. He was excommunicated by the Bishop of Chicago in 1856 and began a new career, first as founder of the Catholic Christian Church and then as a Presbyterian minister. When he encountered difficulties with the Presbytery of Chicago he applied in 1863 for admission, with his Ste-Anne's congregation, to the Synod of the Canada Presbyterian Church and was accepted. The bishops of Québec waged unrelenting war on him, concerned about his oratorical skills and endless attacks on the Catholic Church.

[109]Grace (Wood) Redpath was the wife of Peter Redpath, a well-known Montreal merchant.

[110]Edward Frankland, *Lecture Notes for Chemical Students; Embracing Mineral and Organic Chemistry* (London: J. Van Voorst, 1866).

[111]Royal Botanic Gardens.

Friday of last Week & Monday of this {(tomorrow)} & so I have been roving about a little. I went to Kew with Mr Fisson. We started about two o'clock & went up there by steamboat. The sail took about an hour & a half & as the day was beautiful was very enjoyable. As it was Good Friday the boat was crowded with people out on holiday. You dont' seem to be able to get into the country near London at all, for all the up the river even as far as Kew, both banks are covered with Factories, walls, & houses. There are lots of fine trees however, now just beginning to become green, & many swans floating about on the muddy water.

To get to the gardens after landing at Kew you have to pass over, or beside Kew green, & here hundreds & hundreds of people were enjoying themselves in a rougher way than is allowed in the gardens. Waltzing to crocked fiddles & wheezing acordions, playing at "kiss in the ring", eating perrywinkles, shrimps, & oranges *ad lib*. In the gardens themselves all was propriety & order, not even smoking being allowed. In the various green houses there were many beautiful flowers & the palm-house was magnificent, though I think, not much, if any larger than that which we saw at Edinburgh. Mr Fisson being rather corpulent was nearly melted by the heat in the latter places, & so of course I professed to enjoy it & insisted on going up the winding stairs to the gallery which runs round near the roof.

What I especially admired was the bamboos, their foliage is so light & graceful. There are two museums in the gardens both of which we visited. When we were returning to the boat many people were getting their tea at the so called tea gardens, which are attached to many of the houses, but which I should rather call tea back-yards filled with benches & tables. "Hot water two pence a head". The sail down the river was also very pleasant & the boat not so crowded as on the way up the swans settling themselves to sleep on the mud-banks which the receeding tide had left. I have wondered why they <were> are not often stolen or killed, but it seems that it is 14 years transportation to injure the precious birds.

Yesterday I went down to Woolich & had a short dig at the fossils there & a pleasant <dig> sail up & down, or rather down & up.

April 19. Tuesday

Mamma & Papa have arrived at Londonderry after a very quick & I hope pleasant passage.

I received a telegram from Papa last night which reads as follows. "London derry ten Monday evening all meet Liverpool Tuesday fine or Wednesday morning will telegraph on arrival" & which no doubt should read. "Londonderry ten monday evening, – all well, – Liverpool Tuesday night or Wednesday morning &c". So that I will probably receive another telegram late tonight or tomorrow morning, & I should think they will be here by tomorrow evening. I shall send off this letter this morning if there is any mail (Cunard &c) without waiting for the Canadian mail as usual, so that you may have the earliest news. I daresay Mamma & Papa will arrive in time to write you by that mail themselves.

With very best love

Your affectionate brother

George Dawson to Anna Dawson, London, England, 1 May 1870.

My Dear Anna

I am afraid my first duty will be to apologise for not writing by last mail. But we had such an excursion last Sunday in search of Mr Frasers that I was too tired to write then as usual, & was not able to find time during the week. [&] Then perhaps this letter coming by an intermediate post will be acceptable. Mamma & Papa have now quietly settled down here (20 Halsey St) That is to say if you can call it *quietly* when some great excursion is on hand every day. From most of these I am debarred, as, of course I have to go to the Laboritory as usual every day. But then I see them every morning, & am at home with them in the evenings, or when they go out often accompanying them.

Saturday before last I was at a Reception of the Royal Socy, at Burlington house with Papa. I went to take care of his table of specimens of plants when he was not there. Mamma did not go as ladies were not admitted. There were a great many interesting & curious things to see, which I will not begin to enumerate.

Last Friday I went with Papa & Mamma to a soiree of the Geographical Socy at Willis's rooms. It was quite a gay & festive occasion. The rooms were all hung round with huge maps of newly explored parts of the world & there were tables covered with photographs, sketches, & illustrated books of travel. About 10 o'clock the rooms became pretty full, <&> {who} after wedging round for a few hours left about 12. Last evening we spent very pleasantly dining at Mr Macimillans (the great publisher)[112] He came for us about two o'clock & we spent the afternoon in his garden playing croquet &c. He has a very nice old rambling house, about half way to the Crystal plalace one side completely covered with Ivey, & full of long passages uneven floors &c., but very comfortable I should think to live in. There were only one or two people besides the household, which seems however to be immense all sorts of relatives besides the Macmillans themselves. One of the guests was Mr Lockyer[113] who has made so many great discoveries in the sun lately he seems to be a sort of compound of Dr Hunt[114] & Mr Dankin, & played croquet with the greatest interest & skill.

The course of lectures on physics begin tomorrow, from which time till about the end of June I will have lectures five days every week.

Papa Mamma & I have been this morning to Mr Spurgeons Tabernacle & have heard him preach. He spoke very well & seems to have got over a great deal of the roughness which it was said characterised his sermons at first. He is not at all however preposessing in appearance, rather a low browed, thick-set sort of man.

I suppose some other great excursion will come off this evening & I don't know whether I will go or not. But it it eventuates in Westminster abby perhaps I may

I hope you are getting on all well & comfortably at home & not missing Papa & Mamma very very much. But at any rate it will be a great treat when they return.

Please thank Eva very much for the maple sugar she sent me & Rankine for the stamp.

If the letter is not too heavy when Papas piece is added I will send you a beautiful portrait, (meant for me) who I got cut out by a man on the Woolich boat a few days ago for the small sum of twopence. The one great feature about his portraits were that they would do for anybody.

With Very best love to yourself & love to Rankine Eva & William

Your affectionate

George Dawson to Anna Dawson, London, England, 19 June 1870.

My Dear Anna,

Many thanks for your last letter which though short was welcome & I took the liberty of reading the others also before forwarding them to Scotland.

Papa & Mamma are looking about there for a place to spend the summer. (For Mamma & myself I mean) I dont' yet know how it is to be arranged & whether we are to go with the Bells or the Primroses, or both together, or in short anything about it.

Further than that Papa has spoken of St Andrews & Burntisland as places he thought favourable both as to ordinary sea-side advantages & opportunities for collecting fossils & minerals. Both these places are

[112]Alexander Macmillan (1818-1896), with his brother Daniel, founded and developed the British publishing firm, one of the most important publishing houses in the world.

[113]Sir (Joseph) Norman Lockyer (1836-1920) was a British astronomer who made pioneer observations of solar prominences and speculated widely on many other solar phenomena. In 1869 he also founded the influential science journal *Nature*, which he edited for some fifty years.

[114]Probably Robert Hunt (1807-1887) who was professor of experimental physics at the Royal School of Mines.

quite near to Edinburgh, (within thirty miles) which to my mind is rather a disadvantage as I should prefer to go to some very out of the way place (Hebrides or Shetland Isds for instance?) but then I suppose that would be quite out of the question.

The session will now be ended in two weeks, I will be rather glad when it is over, both because London is getting so unbearably hot & stuffy, & because the examinations will have been got over (for better for worse). With regard to these last I have no very pleasant anticipations. I never have developed the faculty of learning off by heart & a great deal of that is required in the Chemistry & in this hot weather I feel quite done up when I get home & hardly fit to study at all.

The theoretical Chemisty examination comes off on the 29th. & of it I am most afraid. The practical follows the next day, & this depends to a great extent on luck, for you get a substance to analyse, & have seven-hours to do it in. Each <substance> student gets a different substance & though they are supposed to be made as nearly equal as possible, in difficulty it depends a good deal on what substance you get & whether you happen to remember that part of the *tables* well or not. The examination in Physics will I think, though the date is not yet fixed come off the next day (July 1st.). [&] thus the three exams will be on three consecutive days which means only one night to look over the physics.

I have read some more stories in "Men Women & Ghosts" & think them very clever, especially "little Tommy Tucker"

Yesterday afternoon I went for a little while to the Exhibition of the Royal Academy. There are a great number of pictures & some of them very beautiful. Especially some of the landscapes. One of *the* pictures is one called "The Flood" by Millais.[115] It represents a baby in a cradle in the foreground, floating along the stream & stretching up its hands toward the branches of the trees as they pass overhead – quite unconcerned & rather seeming to enjoy it than otherwise. On the corner of the cradle sits a black kitten quite awake to the danger <&> mewing & looking most uncomfortably wet. Beside the cradle floats a big yellow jug, which is most splendidly painted, & in the distance you see the Father & Mother of the child coming in a boat. You are left in uncertainty whether the boat will arrive in time or not as the cradle is not far from the beginning of a sort of rapid.

With best love to yourself & William Rankine & Eva

Your affectionate brother

George Dawson to Anna Dawson, London, England, 26 June 1870.

My Dear Anna,

This will I suppose be my last Sunday in London for some time, at last It will be so if I manage to carry out my present intention & leave next Saturday, either by the 10 o'c morning express or by the night mail.

I am if possible to go direct to Burntisland, (or Burnt-island, or Burntes-land,) where Papa has found a Cottage for the summer.

I intend to mail this letter tomorrow & thus it will reach you by an intermediate mail, & my next will I hope be dated from Scotland. My reason for doing so is that I will of course be very busy about the exams & thus cannot hope to be able to add anything. Besides Papa & Mamma will no doubt write by the Canadian mail. By the way I suppose by the time this gets over you will be at Murray Bay occupying Mrs Henderson's house. How completely the tables will be turned, with somebody else in the *Grande Maison brune*. (As I christened it to make the [...] understand which I meant) You will then have to humbly beg for the use of the ice house & croquet ground. However I hope that better tempered people than ourselves will occupy it. & allow you these privaleges willingly.

I got a letter from Mamma a few days ago in which she told me that the Cornishes were staying in London (or rather at Hammersmith) with the Birds. As I had been there once before with

[115]John Everett Millais (1829-1896) was a member of the Pre-Raphaelite Brotherhood who was extremely talented technically but whose later work was sentimental and facile yet extremely popular.

Mamma I thought I might manage to find my way again, & so went on the expedition this afternoon. I found Prof. C out but Mrs C & Georgie in, both looking very well, & glad to see me. It seemed so curious to see them over here, it makes it look quite a short distance across when so many people you know keep dropping in. She had quite a stock of home news & said which I hope is true that you were getting on very well. She also said that Mr Barnard was ordered off to some other locality in a month or two. She talks of starting for home again about the twenty fifth of August & there is some talk of Mamma going under their escort.

The Weather has been beautifully cool & pleasant for some days back & I hope will continue so till after the exams, at any rate. When I get down to Scotland It may get as hot as it pleases, for hot weather is very pleasant when you are by the sea & have nothing more to do than you please. I think I will presently become if not a "womans rights" man, something like it, for really it is quite painful to think about what small & stupid things womens minds are often occupied. Why should they not, at any rate know as much as other people, for not only would it give them something better to think about, But it would open such boundless fields of pleasure, of a rational kind to them. Why for instance should women not know Geology, Chemistry, &c., for Geology & Chemistry &c. are only in reality the knowledge of things which we meet with every day, & if we know nothing about them but what a dog or cat would know, such as their external form & appearance; pass by without noticing or obtaining any enjoyment from them.

I was very rash in taking two pages, of paper, But you see I have managed to fill up. Do you know how Robert Carpenter is getting on in the west?

Please give R, W, & E *strict* orders not to drown themselves, perhaps the best rule would be that they should write & ask Mamma for leave every time before going out in boats &c.

Give my love to them all & with much to yourself

Believe me your affectionate brother.

George.

P.S. Have you read the Duke of Argull's books called, "Primeval man" & "the reign of law".[116] If not I prescribe them as a dose. They are over here at present but Papa will probably take them back with him. GMD.

[116]George Douglas Campbell, 8th duke of Argyll, *Primeval Man: an Examination of Some Recent Speculations* (London: Strahan, 1869); and *The Reign of Law* (London: Strahan, 1867).

SUMMER 1870, WITH PARENTS IN SCOTLAND

George Dawson to Anna Dawson, Burntisland, Scotland, 6 July 1870.

My Dear Anna,

It always seems of if something prevented me from writing long letters to you.

Though now that I am here I hope to be able to do so, in future. I have made most extensive plans for the summer in the way of Geologising, botanising, microscopising!, sketching &c. [&c.] but do not know how far I will be able to carry them out.

The reason I have not had time to write before, today is that I have been going about Geologising with Papa & a local Burntisland geologist all day by way of getting a notion of the geology of this place before he (Papa) goes away. We had a walk of over two miles there, & of course a like distance back this afternoon, & passed under the cliff where Alexander II.[117] fell over & was killed, in hunting. It is a beautiful cliff & I must take a sketch of it soon. There are lots of beautiful sketches round here & I hope to be able to take some of them & exchange views with you. I have invested in a paint box like yours, but rather smaller, & also got a selection of moist colours for it. All half cakes. It holds about 16 half cakes & Cost altogether about about 20s/. Am not I getting extravagant. I have not yet had any *authentic* accounts of the results of my exams. But <if> one of the fellows has written to me & I have at any rate *passed* in all three subjects & I think within two or three of the top in Chemistry. I hope to hear definitely tomorrow or next day, at latest.

Mamma bids me be sure to tell you to tell Rankine how glad she is to hear that he is first. As William is first two you see I am turning out the stupid one of the family. I hope you will enjoy yourself very much at Murray Bay & have plenty of company & fun.

I cannot begin to tell you anything about this place & our cottage here, but I suppose Mamma has told you all that. <It> I think it will be a very nice place.

We have the sea in front & the railway behind with plenty of trains but not at all noisy.

We have quite a romantic *little* place, with a "lovers seat" up on the point of a little cliff, all included.

Tell W that if he wants specimens of Helix hortenses, or Helix [fornacia][118] (I think) he has only to apply as I think I could gather a bucket full off *"our grounds"* It is only a pity that we cannot *all* spend the summer here. Believing me your very affectionate

[117]The reference should be to Alexander III (1241-1286), King of Scotland, who indeed died here near Kinghorn in a riding accident. Alexander II died on the island of Kerrera on the Scottish west coast.

[118]Snails of the Helicidae family, *Helix* spp.

George Dawson to Anna Dawson, Burntisland, Scotland, 11 August 1870.

My Dear Anna

We were very glad to get your letters this week announcing Papas arrival at Murray Bay. And also letters from Papa Himself dated from Montreal. Yesterday Mamma & I went over to Edinburgh for a trip. We had been waiting for some time for clear weather as we intended ascending Arthurs Seat. The morning was rather misty but it cleared up perfectly by the afternoon. We did a good deal of shopping in the first place, at least Mamma did the shopping & I looked on. We then went for a little while into the museum.[119] It is very nicely arranged & much resembles the South Kensington, in London though of course on a smaller scale & with the addition of many specimens of Natural History. We next had our lunch & then went to Arthurs Seat. First we went all round Salisbury Craigs by the Radical Road,[120] stopping every little while & looking at the view. Then we went up Arthurs Seat itself by a very gradual ascent beginning at St Antony's chapel. It was very hot going up while under shelter of the hill but as soon as we got to the top there was a delicious cool breeze. We stayed there a long time looking at the magnificent view, & Mamma pointed out all the places of interest. The view was really magnificent, the whole Firth of Forth, Edinburgh, the Pentland Hills &c. &c. &c. When we came down we went round by Dudingston Loch & Sampsons ribs, arriving at Gilmore Place about 5 o'clock. We had tea there & started for Burntisland again by the 7.35 train. I suppose Mamma must have already told all this in her letter, but if so please excuse the repitition.

Mamma gave me a beautiful gold chain a day or two ago on my birthday. I like it very much.

Tomorrow is Rankines birthday is it not? Please wish him from me, many happy returns I hope you will give him a party on the ocasion.

Everybody here is in a bustle about the preparations for Janes marriage. More especially at Gilmore place for here we do not have so much of it. We have been out boating several times lately. Once with a man to row to Aberdower, once with David Kemp, & day before yesterday Ellen Maria & myself we went as far as a place called Starley Burn & then rowed back at great leasure singing part of the way it was very jolly the water as calm as a lake, & several other boatloads of people floating at about half a mile apart. Sometimes when one would begin a song the rest would join in & you could hear them quite clearly across the water. (N.B. I dont' mean at Murray Bay)

With best love to Papa, W, R, & E Mrs Carpenter & yourself

Yours ever affectionately

[119]Now the Royal Scottish Museum.

[120]Queen's Drive.

SECOND YEAR AT THE ROYAL SCHOOL OF MINES, 1870-71

George Dawson to Anna Dawson, London, England, 12 October 1870.

My Dear Anna,

London again, & work, but as yet not very hard work. In fact I feel rather lazy at not having more to do. The mineralogy I find does not begin till the 8th. of next month. At present I therefore have chemistry lecture from 10 till 11, then time to read in the library &c. till 2 & then natural history till 3.

The natural history notes take most of the evening to write up & as I am writing this after having written my notes for today, you must excuse everything.

I have heard nothing more from Maggie Frier & so suppose that she has left London before I came up. I found before leaving Scotland that I had promised several more photographs than I had. Mamma having taken away most of my dozen with her. I would have had to pay 6s. for half a dozen of the same & so thought it better to try a dozen, for the same price at a cheaper place. This I did & send a copy (the first) by this mail. I think it is more of a success than the former, but you must tell me what you think of it. The rest I expect from Edinburgh every day.

Oct 13

I have just this morning received my this weeks letter. It being rather late. When is your great houseful going to disperse?

Your affectionate brother

George

{I find I have not got an envelope G.M.D.}

George Dawson to Anna Dawson, London, England, 23 October 1870.

My Dear Anna

Here I sit down again to write, & unfortunately with nothing to write about. Things have got into running order, & run on very smoothly with out any incidents worthy of mention. You should not scold me for writing such commonplace letters, for you know I cannot invent news when there is none to tell, & neither am I one of these impressable people who can write gushing letters, full of affection, or rather expressions of it. You must not think that I do not feel all this, even though I dont' express it in every letter. My letters to you, & those at home, are not like those which might be written to some distant friend, for they are characterised by a free & easy carelesness which I would not use towards others, & I am afraid that this carelesness even extends in many instances to spelling! Slips of that kind please correct, & keep on correcting till they dissapear.

Mamma wrote lately about music, & said that you thought it would be so much cheaper to get it here. I have got a catalogue of the

Musical Boquet, & Ellen is to write me & give me the names of some pieces she considers pretty. I have asked Maria Cleghorn to do the same, & when I get these I will choose some of them, & send them out to you on experiment. The Musical Boquet is not, as Mamma seems to think a place where second hand music, used by professional {people} &c. is <used> kept, but a series of standard & other pieces, published at a very low price. I saw a lot of second hand music at a shop near here the other day & went in & looked at it. It was for the price of 3d a piece. But mostly very trashy or effemeral. I got a couple of pieces however. The best way will be to send you the catalogue after getting the first lot, & letting you choose for yourself, send me the names, & I will get them for you.

My days at present are mostly spent at the library at Jermyn St. Reading & looking over notes &c. The writing I do in the evenings after dinner. It usually takes me the best part of the evening to, reconstruct & write out my Nat Hist notes. I will have to cut them shorter after the Mineralogy begins.

I have been trying all sorts of places to get my lunch, in the middle of the day, & have found several very decent places for that purpose. I will leave a little space at the foot here in case I find anything more to say before I mail.

Oct 27.

Many thanks dear Anna for your pretty little sketch, of the sunrise. I like it very much, & think it very well done, though I do not at all profess to be a judge. I like the bredth of style however, & sketchy appearance, more than any others of yours I have seen. I only wish I could get it myself. It is so hard, in copying from nature to avoid getting into a cramped finicky sort of way. By next mail, I will look out one of mine & send you in return.

I really did but very few during the summer, & most of them are, & must remain for some time unfinished. & what are done I like less every time I look at them, they are so little like really good artistic sketches, & so very weak & washy

Believe me your affect brother

George Dawson to Anna Dawson, London, England, 20 November 1870.

My Dear Anna

The socks which you sent, & which arrived quite safely, I have tried for several nights, & find them to be remarkably comfortable inventions.

I am already beginning to look forward to the Christmas holidays, as the next event of any importance, I hardly think it likely that I will go to Mrs Reids, for besides the expense of the journey which would be considerable, I think I will have a good deal of profitable work in looking over notes, &c. The holidays too are very short, at the utmost ten days. Both Natural History & Mineralogy extend past Xmas & then Geology begins, which will need considerable time to do it justice, besides having the other subjects to keep up.

With regard to the very solemn question in your last, about the advisability of a woman knowing something by which to earn her living. I hardly know how to answer, but if in your case it means improving yourself in painting, then by all means *yes*. I wonder you don't try some oil-painting, I think it must be very nice work. Also painting figures, cattle &c. is a very interesting kind of painting I should think.

You will be "snowed up" no doubt by the time this reaches you, & all made hard & clean for the winter. We here are having any ammount of cold chilly weather, & rain. Far colder than Canadian winter to the feel. I had a letter from Mrs Dunscombe a few days ago, she has a little son of two months old, which she has very properly, & with total disregard for parental cognomens, called Sydeney Herbert, or Herbert Sydney (I forget which) meerely because she thought them pretty names.

The Eastern question[121] has turned up again, but I dont' think will come to war, but either end in a conference or silent dropping of

[121]A reference to the issues surrounding what became the Russo-Turkish war which, contrary to George's prediction, lasted until March 1878 when Russian forces defeated the Turks.

the question by Russia. Papers of *all* shades of politics here are united on the subject, & approve of Lord Granvilles dispach.[122]

At any rate by all accounts it would not be hard, with Austria & Turkey to give the Russians another Crimean lesson. The Turkish fleet alone could now bottle up every fort in the black Sea, & take & keep the now unfortafied Crimea.

One <rather> disadvantage of living as I now do at Mrs Guests, is that a great deal of time is wasted in talking &c. & the only remedy is to sit up rather late to finish your work.

I usually now go in for a roal & butter & a cup of Coffee, for lunch & having found several rather good places oscilate between them. The cost is 6d.

Novr 24

Thanks for your letter which arrived in good time day before yesterday.

Nothing new to report,

With much love to yourself & all

Believe me your ever

affectionate brother

George Dawson to Anna Dawson, London, England, 15 January 1871.

My Dear Anna,

Last night I was at Westminster Hall, & saw the Annual inspection & presentation of prizes to the London Scottish Volunteers. George Primrose belongs to the corps, & sent me a ticket.

It was rather interesting, & speeches were made by Lord Elcloe, the <chair> colonel & also by Sir William <Annesley> Mansfield.[123] The latter however was not well heard, so that I cannot tell you what he said in his very interesting speech, till I see it in the papers tomorrow. A great many prizes were also given, for shooting &c. The corps is a strong one mustering 749. Their uniform is rather pretty consisting of a kilt, plaid &c. of a grey colour. This afternoon I went to the service at Westminster Abby, & being very early managed to get a good place, for once, inside the rails, & heard & saw well. Dean Stanley[124] was the speaker, & his text was the last few words of the Lords prayer. "For thine is the Kingdom &c". He said they were not originally part of it, but were well known to have been added, though at what time was uncertain. From this he proceeded to argue that the words of scripture should not be taken too literally & that parts such as this which had been introduced by the universal consent & aprobation of the church posessed almost equal authority. I dont' care much for his preaching he seems to be such a platitudinisary old fellow.

Jan 19.

Many thanks for your nice long letter from Toronto. I am glad that you have been enjoying yourself so much there. I only wish I was in circumstances to collect so much news, & write such long letters, but unfortunately everything goes on much the same, day after day. As you will no doubt be home again now, please tell Mamma that my hyacynths she asks about, are getting on passably well but very slowly. The cold weather we had lately seemed to put them back, but no doubt now that we have returned to wind & rain, that they will begin to grow again. You ask me to tell you all about the various places I was out at, during the holidays. I think I did so at the time. I called at the Raes the other day, & Mrs R desired to be remembered to Mamma & to thank her for sending Papas lecture. They seem to be very nice people.

The double number of letters which arrived this morning have taken me so long to read that I have very little time left to write, before

[122]George Leveson-Gower, 2d earl of Granville (1815-1891), was then British foreign secretary. Granville's protest of the Russian denunciation of the Black Sea clauses of the Treaty of Paris, however, eventually proved ineffectual.

[123]Sir William Rose Mansfield (1819-1876) was a renowned British soldier who had distinguished himself in India.

[124]Arthur Penrhyn Stanley (1815-1881) was a well-known British churchman and dean of Westminster from 1864 to 1881, who, nonetheless, obviously failed to impress George.

I must go out & get my lunch preparatory to Mineralogy & Natural History lectures There please excuse more at this time & believe me your affectionate brother

George Dawson to Anna Dawson, London, England, 2 February 1871.

Dear Anna

I was very glad to get your long interesting letters, by Cunard mail yesterday morning, as I had nothing from you last week & Mamma said she had heard nothing since she wrote before. I am very glad you have been enjoying yourself at Toronto so much, & hope you will not find it dull at home after so much excitement, religious & otherwise.

I sometimes see the "Times" when I go out for lunch, & while the siege of Paris lasted,[125] it was quite a curiosity. Washbourne,[126] the U.S. minister was the only embassidor who stayed in the city, & his weekly bundle of dispatches were the only communications allowed to pass in. As the "Times" was included in these, French refugees & others having friends in Paris, hit upon the plan of advertising whatever they wished to say to their besieged friends in the Times, & generally adding a request that, Mr Washbourne would <let them> forward the intelligence to them. Lately the whole first page of the Times was covered with such advertisements, so that it must have been quite a task to deliver the messages.

You say I sometimes only send you little scraps, which is quite true, but then you must remember that you only have to write one letter, while I have to answer two or three every mail, which not only takes much more time, but, requires an amazing fund of news. Please excuse this "little scrap," as I have already written so much, & with very best love both to yourself personally & all at home.

Believe me

Ever your affectionate Brother

{(P.S. – Mc.Amber)}

George Dawson to Anna Dawson, London, England, 9 March 1871.

My Dear Anna

I have just finished reading my letters which arrived this morning. I do not quite understand the purpose of your tract meetings though you take such trouble to explain it, what do you do with them when they are bound up into books,? & what is the use of sticking pictures &c. into them unless for children?

Last Saturday evening I had a visit from Mr D.W. Kemp. Ellens husband, he had been called up to London quite unexpectedly by business. I arranged to meet him at Farringdon St Station the next morning at 10. I accomplished this, & we went together to St Pauls where the service begins at 10.30. It was a little long & tedious, but worth going to for once. Afterwards we went for a walk in Hyde Park & Kensington gardens. Prof Ramsey[127] had been kind enough to ask me to dine with him again last Sunday, & as I went there in the evening my day was pretty fully occupied. Last evening I went up to Kings Cross Station by underground to see him on his way back to Scotland.

The notice which we have been looking for about the mineralogy examination has at last been put up. The examination is to take place on the 31 of this month, leaving now just three weeks to prepare. I

[125]The siege of Paris which began on 19 September 1870 had lasted until 28 January 1871, when the French surrendered to the Germans.

[126]Elihu Benjamin Washburne (1816-1887) was a prominent Civil War era American politician who was Minister to France during the two terms of Ulysses S. Grant.

[127]Sir Andrew Crombie Ramsay (1814-1891) was director-general of the British Geological Survey. One of the most distinguished nineteenth-century geologists, Ramsay had an enviable reputation as a field geologist and innovative theoretician.

feel quite afraid of it, there is such a terrible mass of minerals to learn up, & so much pure memory required. Yesterday I was at the British Museum by 10 o'clock & met two of other fellows there by appointment, & we spent 3 hours going over the minerals, some of which are really splendid. The only drawback is that they are arranged chemically after Berzelius's method[128] <at the> & To find the minerals & ores, of any one metal, or base you have to look in perhaps a dozen different cases, & as the collection occupies 4 long rooms, you have a great deal of walking to do.

I dont' quite remember the position of the Macphersons house at Cacouna, which you have taken, but imagine, I dont' know how correctly that it lies at some distance back from the road, on the right hand going towards the village, & just before you go up a little hill in the road, the top of which has woods of some kind or other on it. Perhaps you will not be able to tell me if my description is correct or not yet, but at any rate you will be able to do so when you get down there. Cacouna though it is rather flat is a very pretty place, & I seem to have a greater affection for it than either Murray Bay or Tadousac, though all of these places seem like second homes.

I forgot in my last letter to ask you to thank Nina for the little note she wrote to me, & now I have in addition to acknowledge one from Sophy Brown, please say that as soon as I can I will try to answer them. I had a letter the other day from Mrs Dunscombe if you remember who that is. She has now a little son, <which> whom she has Called ____. ____. (Two very nice names but I forget them) not as she says after anybody, but simply because they are pretty. I intend to congratulate her on so far breaking through president. I have lately got into the habit of sitting up quite late at night, but am going to try & reform, for I find that when I get to bed on the right side of twelve I am much fresher in the mornings. I have a little cold at present but I think it is getting better. I fancy I caught it in St Pauls last Sunday, for that building is so large that, although tight enough outside, A regular system of trade winds are set up within the church itself. With best love to yourself, Mamma, & all at home, believe me. Your's

George Dawson to Anna Dawson, London, England, 19 March 1871.

My Dear Anna

You have no doubt heard that there is to be an exhibition here this summer. It is only lately that I have been able to understand its object. It is to be a permanent International exhibition. That is to say that each year only a certain class of articles will be allowed entry, & that in about ten years or so, the round will have been completed & the classes of objects shown in this, the first exhibition shown again.[129] This year the exhibition is to open I believe in May & to consist of machinary in motion, porcelain & pottery, & something else which I forget. If it is successful, as it no doubt will be, it will be well worth going to see. At the opening of the exhibiton the new Albert Hall of Arts & Sciences,[130] is also to be opened, with great Ceremony. It is really a huge building, & when you see it from the park quite towers above the neighbouring houses. It has been built by shareholders, each getting, according to the amount of his investment, a number of seats. It is supposed to be built on the most scientific principles with regard to sound, & will be used for concerts, picture exhibitions & so on.

I am afraid you will consider this much too grave & serious, as you say my letters are so generally, but really it is very difficult to write entertaining letters, for in writing you have the conversation all to

[128]Jacob Berzelius (1779-1848) was a Swedish chemist who first presented a mineral classification based upon chemical composition.

[129]The exhibition referred to by Dawson which began in 1870, was indeed confined to one or two special descriptions of produce or manufactures. These exhibitions continued until 1874 but failed to attract a very large public attendance and were abandoned.

[130]Royal Albert Hall.

yourself, which is very tiresome. I much prefer indeed someone who will continue talking in a not irrational manner, with only such slight encouragement as is afforded by "yes" & other such monosyllabic words. This description answers to at least one of your friends, – of whom by the way I have not heard anything of for ever so long – I daresay you will know which.

Ideas & those sort of things cool before they have been carried three thousand miles, & I have no doubt, that if we could have a branch of the Atlantic Cable laid on, we would get along in that respect much better. I am principally occupied just now by preparing for the mineralogy examination which comes off on Friday week, & will no doubt be all over before this reaches you. I have no very pleasant anticipations concerning it, but will of course try to do the best I can. As we have been given, in the lectures about 300 minerals & have say 10 facts to remember about each, this makes altogether about 3000 facts to remember for the examination, besides general reading on the subject, & acquiring a facility in distinguishing specimens. On the fifteenth of next month comes the drawing exam. which is if anything more to be dreaded, as I do not take at all kindly to the subject. I hope to get through however & will then be able to forget it as soon as I like.

I was interested in the account you gave in your last of the newsboys Sunday School, but hardly think it the sort of work you should undertake. I should fancy there are many much less martyrising lines which would be quite as useful.

This letter seems to be written in the *pot au feu* style – please tell me how you like it

Yours with best love

George

March 23

I have just got this weeks letters, for which thanks, also Goldwin Smiths[131] Address & various newspapers. I enclosed a cutting from the "Echo"[132] which please show to Sophy Browne as it is on the subject which she started in her note [...]. Have also got W's P.O. order.

George Dawson to Anna Dawson, London, England, 3 April 1871.

My Dear Anna

I have to thank you again for an interesting & long letter by last mail, & was quite astonished to hear that Mrs Baynes had arrived at the stage of being Called Grandmother. You mentioned some time ago that O'Hara had gone into Molsons Bank, & seemed to think that it was a very suitable place for him. Sophy seems to say very much the same thing in her note. Why has he left college? & why do you think his abilities so small as only to suit him for such a post? I thought he was going in for a medical course.

Thinking I was entitled to some dissipation after the mineralogy examination, I yesterday started at the early hour of nine, for Mortlake to see the Oxford & Cambridge boat Race. Last year I saw it about the middle at Hammersmith, so this time I thought I would see the finish, which takes place just at Mortlake. I went by the Underground Railway to Victoria, changed there & went to Clapham Junction. There I began to get into the crush of people & had some trouble in getting my ticket, especially as it was necessary to go to one station for the ticket, & afterwards to another to wait for the train. I got one however, & after a few minutes managed to squeeze into a train. At Mortlake though so early the place was all alive with people, & crowds pressing toward the river, on foot & in carriages. The banks were densely covered with people, but I managed to get on a barge, of which a great many were moored along the course. The race depended on the tide, & so the time is always slightly uncertain. There was a

[131]Goldwin Smith (1823-1910) was a Canadian historian and journalist who had settled in Toronto in 1871 and wrote extensively on Canadian and international affairs.

[132]A London daily evening newspaper.

band, just opposite where I was, & so between watching the people, & listening to the music the time passed quickly enough. In about half an hour, there was a murmering sound heard down the river, & gradually this became louder & louder till the boats appeared & the sound became a regular roar. Cambridge was ahead, & both crews went fast in beautiful style. Though the Oxonians seemed not to be quite in such good form as the Cambridge men, & to throw up the water a little with their oars. You will have heard long before this that Cambridge won by about two boat lengths. The instant the boats were past, hundreds of skiffs & boats shot out from the banks & among the barges, & the whole river was covered with them bobbing about in the swell raised by the <judges> Umpires steamboat, & those belonging to Oxford & Cambridge, & holding their respective friends. As they were all covered with flags, & bearing the dark or light blue, it was a very pretty sight. Getting back to town again was rather more difficult on the whole than going out, as all the people of course wanted to do so at the same time. The Train to Clapham waited about three quarters of an hour at Barnes, with another waiting in front, & still another behind.

Tomorrow is the Census & we have just been filling up that very inquisitive paper. I know you are to have one in Canada, but cannot remember whether it is on the same day as the one here.

April 6

Many thanks for your letter just received. Easter Holidays are now supposed to have commenced, but I have come up to Jermyn St today as the Library is open, as it will reduce the length of those horrible do-nothing days.

Your affectionate brother

George Dawson to Anna Dawson, London, England, 25 June 1871.

My Dear Anna

I have to thank you again for a nice interesting letter. I really believe that with so much company at home, you must enjoy yourself more than you make out & that it must be an agreeable change.

I have only Palaeontology to work at now, but find it quite enough. I shall have had about a months steady work at it by the time the exam comes on, & if I do not pass this time, I think I will hardly feel inclined to go in for it again. My occupation is about as follows. Get to Jermyn St by 10 o'clock. Read palaeont, or grind up the tables of ranges. Then go up to the museum at 12 & have an hours work at the cases. Parker[133] goes with me, & we are working the subject together, though he is much better at it than I as this is his second year. Then we go out to lunch & often improve the time by holding a little examination. In the afternoon we hold an exam. for an hour, have another hour at the cases, & fill in the extra time with reading &c. As we walk homeward <we> part of the way together we usually then have another short exam.

In the Evening I read over the notes & grind at the discouraging "tables of ranges" &c.

There is a series of lectures going on <on> in the *School of Mines* Theatre on Saturday afternoons. They are in aid of some working mans club. They have got several good names down as lecturers. Amongst others Dean Stanley, & Hughes, the author of Tom "Brownes school days"[134] Yesterday it was Miss Emily Faithfull,[135] the subject of her lecture being "The best society." I went in for a few minutes

[133]A student at the Royal School of Mines, T.J. Parker was the son of comparative anatomist W.K. Parker.

[134]Thomas Hughes (1822-1896) was a well-known British author whose huge bestseller, *Tom Brown's School Days*, was a portrait of the figure of the high-minded sportmanlike English upper-class boy.

[135]Emily Faithfull (1835-1895) was an English philanthropist who took a great interest in the conditions of working women. She lectured widely and in 1863 founded then edited the monthly *The Victoria Magazine*, which advocated the claims of women to remunerative employment.

toward the end just to see what she looked like. She seemed to lecture pretty well, but was not by any means fascinating in appearance, being rather old & quite too fat.

You seem to think that I have a sort of preference for this country, & will like to remain here after I am through at the School of Mines. This is not at all the case I would a hundred times sooner live in any part of Canada. I dont' like the climate here at all, <&> nor do I think that it agrees with me any better, or so well as the Canadian. I think we have had rain every day for a month now, & today it is miserably cold & quite makes one long for a great coat again. As somebody very well said. The summer has set in with its usual severity. It just strikes me that this letter will not reach you till you will in all probability comfortably settled for the holidays at Cacouna. How I wish I could go there too. But I must not allow myself to begin thinking about home as yet.

Your affectionate Brother

George.

June 28 Thanks for your letter just received. I note what you say about my not having written to Nina & feel very much ashamed of myself. I will try & write next mail. Please also make my excuses to William Rankine & Eva to whom I have not written for some time & say that I will try to do better in future.

Yours.

George Dawson to Anna Dawson, London, England, 9 July 1871.

My Dear Anna

I had hoped to leave London before this, but Prof Ramsay wrote to Mr Ward[136] in Cumberland, on Friday last & thought of course that it would be better that I should wait till he heard his exact whereabouts &c. on Monday (Tomorrow)

The results of the year were put up yesterday afternoon, & I am very agruably surprised by the result. I copied out the entire list, & send it to you, as you may know some of the others at least by name. You will see that I am first in Mineralogy & also in Geology, the latter position entitling me to the Directors Medal & a prize of books, the books you can choose as you like providing only that they are scientific The value of the prize is about £20. I have also got a first class in Mechanical Drawing & the same position in Palaeontology about getting through which I feared so much. Parker & I worked the latter subject together, & very fortunately we have come out bracketed as equal at the top of the first class. As having the greatest number of marks for the year I have also come into a scholarship, which is granted by the Prince of Wales, as Duke of Cornwall; & is worth £60. Thirty pounds being payable this year & the other 30 sometime next, on the condition that you attend the lectures regularly & pass the exams with credit. I had no idea of all this & did not even know that the scholarship depended on the second years work. I had hoped to try for the Geology medal but honestly thought that I had spoiled my chance in the examination. I was very much surprised therefore on going up yesterday to be received with congratulations on all sides. Prof Ramsay told me that I got 99 out of the 100 marks, & that the next man got 98 so that it was a very close shave. Had he been first however I would still have got the medal as Elwes had not gone in for the exams in any other subject & was therefore ineligable. I got my medal yesterday but am to take it up again on Monday & leave it with Mr Reeks to get my name engraved on it. It has Sir R. Murchison's[137] head on one side, & a trophy of fossils & Hammers on the other, the whole being surrounded by a wreath of graftolites. It is Bronze like all the other School of Mines medals.

With all Love

Your affectionate Brother

[136]James Clifton Ward (1843-1880) was a prominent geologist who had also trained at the Royal School of Mines and was then attached to the Geological Survey in the Lake District.

[137]Sir Roderick Impey Murchison (1797-1871) was a pre-eminent British geologist who had been appointed director general of the Geological Survey and director of the Royal School of Mines in 1855.

A SUMMER OF FIELD WORK IN THE ENGLISH LAKE DISTRICT

George Dawson to Anna Dawson, Keswick, England, 19 July 1871.

My Dear Anna

You must excuse a <long> short letter as I am going out in about twenty minutes & have only time to write a few words before I go.

I wish you were here to employ yourself in taking sketches. Every way you look you see a beautiful picture, every hill has a magnificent view, & the valleys which run up among the mountains as they turn & bend would give hundreds of pictures. The hillsides are generally bare, or only very scantily wooded. The vallies have fields & farms & are traversed by narrow winding lanes which are generally embowered in trees, & lined by great bushy overgrown hedges. Yesterday we went up Wythburn. It is a beautifull brook & comes down a deep valley in a series of cascades with deep black pools between. The mountains are generally somewhat broad & level on top, & these "fells" consist for the greater part of swampy bogs full of wiry grass & having little pools. From these the brooks are fed & run down on every side. The mountains are only used for sheep pasture, & you seldom meet any living thing on them, but sheep. I have had as yet hardly any time for sketching but hope to make a few little drawings before I leave.

With love

Your affectionate brother

George Dawson to Anna Dawson, Keswick, England, 20 July 1871.

My Dear Anna

I think it is your week for a letter but am not quite sure, at any rate you shall have the benefit of the doubt.

Things go on very much as they did when I last wrote, except that during the past week the weather has been even wetter than usual & has prevented Mr Ward from having a whole days work occasionally. It is always a bother in the morning when it looks threatening to decide whether to cut lunch or not. The decision generally is to go out & chance it & this very often ends in being rained on the best part of the day. Mr Wards Parents are coming in a few weeks from London to settle here, & he has got a house ready for them. One of his sisters is already here, & married to a Dr Knight. They have been kind enough to ask me twice to tea since I have been here. One day lately we were up at the head of Borrowdale. We started about 10 o'clock & walked up by a very beautiful road to a place called Watendlath It is a village of about six houses which cluster together under some great pines, just at the edge of a little lake called Watendlath Tarn. The road is very beautifull part of the way through woods & at one place you come out on the edge of a cliff & have a magnificent panoramic view of Derwent Water & all the surrounding country. We examined the course of a trap dyke[138] there for some time, got rained on once or twice, & after having lunch, walked on again up Borrowdale. I saw the far famed plumbago[139] mine, or rather the place where it was, the view consisting of several heaps of rubbish on a hill slope. There <are> is a grove of very old yews (said to be about the oldest in England) not far from the Plumbago mine. They are splendid trees, & though I do not know the measurement of their trunks they must be very large. I say, a grove, but in reality there are only 3 or 4 trees, which shadow a space large enough for a grove

Seathwaite the village (about 9 houses) nearest to the Plumbago mine is said to be the wettest place in Britain. I can quite believe it, though, wonderful to say it did not rain while we were there. We noticed a bridge across the stream which looked somewhat askew, & have heard since that it <...> {has been} carried away by the water during a wet night. [illustration in letter of bridge]

As I hope to leave for Edinburgh on Tueseday Aug 1. I will probably write further from there before closing. I wrote to Aunt to say I was coming & received by return a very kind invitation.

Believe me

Dear Anna

Your affectionate Brother

George Dawson to Anna Dawson, Edinburgh, Scotland, 9 August 1871.

My Dear Anna

I have been hard at work for the last week attending the meetings of the British Association[140] This afternoon they are over unless I count the geological excursion – on which I intend to go – or other of tomorrows excursions.

Last night I was out at a soiree in the Museum of Science & art[141] – It was a very pleasant evening but with all very hot, even though from the size of the room they were not unpleasantly crowded. We had some very good music from the band of the Scotch Greys, & I was so fortunate as to meet a good many people I knew.

Amongst them Miss Taylor who came over in the same ship with Mamma last time.

I dined with Col Lyell & family last Sunday evening. They are here for the B.A. Meeting & have very good rooms in Princes Street.

On Saturday last the Geological Section was adjourned at one o'clock to Arthurs Seat & Prof Geikie[142] explained the structure of the hill very well. It was a very windy day but fortunately fine. It was amusing to see the people being blown about. At the top it was quite a gale, & very hard to hear Geikie at all. The grass was drie & slippery & in coming down many surprised themselves by their own velocity.

I have just come in from an "evangelistic meeting" held in the Music Hall. I did not care particularly to go especially after all the meetings of the day but as Ellen seemed very anxious for me to do so I went. It was in the usual style. Afterwards we went along Princes Street & saw the Illuminations & decorations put up on account of Sir Walter Scotts Centenary. The streets were perfectly packed with people & we managed to get along only with the greatest difficulty. There was no general illumination – but many of the hotels &c. were very nicely lighted. During the day most of the streets were hung with flags, & the town even quite late in the evening full of country people.

[138] A now obsolete name for a compact, fine-grained igneous rock in a sheet-like body which cuts across the bedding of the host rock.

[139] Or graphite.

[140] British Association for the Advancement of Science.

[141] Royal Scottish Museum.

[142] Sir Archibald Geikie (1835-1924) was a Scottish geologist and professor who was later director general of the Geological Survey.

I send by this mail the newspapers containing Association reports, & hope you will get them all safely.

Perhaps Mamma [might] like to know that Mrs Reid has another baby, a daughter.

Yours with love

George Dawson to Anna Dawson, Edinburgh, Scotland, 23 August 1871.

My Dear Anna

You will see that I am still in Edinburgh. I had meant to return to Cumberland before this but on writing to Mr Ward found that the arrival of his Parents from London had been delayed beyond his expectations & that he was in quite a transition state I have therefore thought it better to stay here for a day or two. Especially as Mr Wards father is now staying at his lodgings & other accomodation at Keswick can not be found except at a hotel now that the season is at its height. I therefore intend to start for Keswick next Monday (Aug 28) & will then go into the rooms vacated by Mr Ward.

Last Saturday after consulting my geological map I decided to take a Walk from Granton to S. Queensferry. Ellen & Louise went with me. We went to Granton by train & then walked along the shore to Crammond. Here there is a little river to cross (Almond Water) & we were not sure whether we could get over or not. However fortunately Lord ___ at whose property we wished to get was from home & so the ferryman made no difficulty about taking us over. Then we walked along a very beautiful path along the seashore till we came to Barnbeugles Castle a beautiful ivy-clad ruin though not very old I should think from the style of the architecture. They tell I do not know how truly that after dinner one day, the wife of the former Lord escaped in a boat from the castle steps with some young gentleman, & that since then the castle has been allowed to fall into ruins & a new mansion built in a different part of the grounds. We did not stay long at the castle but went onward by a very pretty walk through woods & near the shore to S. Queensferry. Here we were fortunate enough to just catch the 3 o'c train & got back to Edinburgh after a very pleasant <walk> day. I have not had this weeks letters yet but hope for them tomorrow morning. I have not had a letter from Papa or William for a long time. I suppose they are almost beyond post offices in P.E.I. I got last week besides the Letter from Dora which you enclosed a note from Nina. I hope you will thank them both from me, for their kindness in writing. If Annie comes to Montreal as you said she thought of doing I hope you will give me some [...] Mrs Baynes will no doubt now be back. I had meant to write her a note before she sailed but unfortunately forgot the day till too late.

I have been up this evening to tea at Maggies & afterwards for a walk to a place called Gorgie & back by the Corstorphine road. Little Edwin is running about manfully now & speaking quite well. He is very clever & bright. His photograph has just been taken, & Maggie is to send one out as soon as she gets some copies. Speaking of photographs when down at Musselburgh Marian wanted very much to have one of Evas, the one in which she is standing in a white dress. I promised to ask if you had any spare copies at home if you have please send me one for her. Hoping that this will find you comfortably settled at home again & much better of your summer

Believe me

Your loving brother

George

P.S. Aunt & all here desire to be kindly remembered. *GMD.*

George Dawson to Anna Dawson, Keswick, England, 3 September 1871.

{Have just received this weeks letters. Stamps enclosed for William please thank him for the ones he sent. *GMD*}

My Dear Anna

Today has been thoroughly wet, & a wet Sunday is always unpleasant. I went to church in the morning & heard a sermon from no less a personage than Mr Ryle,[143] known as the author of Tracts for the times &c. &c. The sermon was very fair but nothing particular. He speaks loud but not very distinctly. In fact his words seem somehow to echoe in his mouth before they get out & so become confused. In the afternoon I fortunately found a copy of "Kitty Travylliows Diary" & so reading that has kept me out of mischief for the rest of the day. I am staying now alone at Mr Wards old lodgings, & find it a considerable change to be so much by myself after being with so many kind friends in Edinburgh.

The Wards are now settling in their new house. I cant' say settled as they have not yet managed to get a servant, but have to trust to precarious charwomen & are otherwise in a chaotic state. The work at preasant lies up Borrowdale. We start by an omnibus at 9.45 & go by it about 3 miles to the Borrowdale hotel. From here we walk up Borrowdale 4 or 5 miles to the point where work begins. Yesterday we had rather a longer & more eventful day than usual. We climbed up by a brook called unpoetically Sour Milk Gill, into the coombe[144] in which it takes its rise. This coombe is walled in on every side but one by an amphitheatre of mountains steep & cliffy at their tops, surrounded at their feet by long slopes of debris & huge masses of fallen rocks. The bottom of the coombe is flat & marshy & the little stream winds through it. The water is beautifully clear & in every little pool we saw {small} trout rushing about & getting under stones as they caught sight of us. After working there a while & having lunch, we climbed up one of the sides of the coombe & got to the very summit of a high ridge of mountains. It was very clear & we had a magnificent view. The height was about 2400 feet. Looking down to the East we Saw into Borrowdale deep & shut in by mountains, & the fields lying flat & green along the river. Further away the range of which Helvellyn formed part appeared overlooking the tops of several mountain ridges & closing the horizon. More to the North the Pennine Mts forming the backbone of England were clearly visible through a broad gap. Northward Saddleback & Skiddaw stood bold & clear. Further west in a long narrow valley lay Buttermere & Crummock water shining like mirrors, & seperated from each other by a flat alluvial strip of land. Following this Valley you could see it widening out into the plain. Beyond that the Solway Firth shining & on the very horizon the Scottish hills[145] of its further shore. The sea was visible through two [other] gaps in the mountains one nearly West & looking up the valley in which the long lake of Ennedale[146] lies. The other a notch with the bold cliffs of Great Gable at one side showed a part of the coast which must have been near the Mouth of Morecombe Bay.[147]

The sun was pretty low & the colouring of the landscape splendid. The mountains full of deep indigo shadows & purple lights. One might have looked all day, but we had not a minute to spare & so had to hurry on.

If what Mamma says in her last letter about your coming to London to study painting is anything more than a joke, of course I need not say how delighted I would be. I do not know what arrangements would be proposed. I feel bound to go back to Mrs Guests for at least the first three months, & of course a students boarding house is not a place for you to come to, if indeed – which is unlikely – there would be any room. The cheapest & best way where there is more

[143]John Charles Ryle (1816-1900), an Evangelical Anglican known for his many religious writings, later became Bishop of Liverpool.

[144]Or coom, a dome-shaped hill or ridge.

[145]Southern Uplands.

[146]Ennerdale Water.

[147]Moricambe Bay.

than one person is to lodge – meerely paying so much per week for Rooms & attendance, & keeping yourself as you like.

Goodbye for the present & hoping my *descriptive* letter will not bore you

Believe me yours

George Dawson to Anna Dawson, Keswick, England, 17 September 1871.

My Dear Anna

Your long & acceptable letter arrived last week by the Canadian mail, insead of the Cunard. No doubt it had been too late for the latter.

Last week has been exceedingly fine & we have not been obliged to stay in a single day. Fortunately we had not any very long trails & so the fatigue did not accumulate. I think I get to admire the scenery more & more, always stumbling on some new beautiful view or picture just waiting to be painted. I often wish I had time to sit down & try my hand under the guidance of some competent teacher such as yourself, but as we are always on the move I have not even the chance of trying alone. I have a little sketch book which I got in Edinburgh which I carry in my pocket & I sometimes get the chance to make a little pencil sketch or outline. This chance generally comes after lunch when we usually take a few minutes rest, so that the subject of the sketch depends altogether on where we sit down for that purpose (lunch). If I get leaves enough of my book covered to make it worth while I will send it to you by post for your inspection. Artists are very numerous here at present & one is constantly coming on them along the roadside or perched on top of rocks. They generally do not stray far from the frequented paths, but one day we were climbing up the stony bed of a brook (which ran in a gully overshadowed by trees & very much reminded me of the Grande Ruisseau at Murray Bay) enduring a gentle rain which we hoped would pass off, when what should we see on turning a corner but an indefatigable artist using a rock for a table, & patiently painting away in front of a waterfall under the shelter of a ponderous gingham? Coming out of church this morning I was astonished to hear a lady speaking to me, & still more so when on turning round I saw Mrs Thomson & her daughter Dr Rae's relations in law. Well known to Mamma. They have been staying here for some time but we never chanced to meet, I suppose because I was never about the town during the day.

I received the "Leaflet" sent in lieu of the tract quite safely, have read it, & think the author would have done a deal better to stick to prose composition.

Excuse this unsatisfactory letter & believe me as ever your affectionate brother

THIRD YEAR AT THE ROYAL SCHOOL OF MINES, 1871-72

George Dawson to Anna Dawson, London, England, 8 October 1871.

My Dear Anna

I must appologise for my very scanty letters of last week. I have also got quite out of my routine & forget to whom I have & have not written lately, though I fancy were my accounts balanced the number in the latter category would much preponderate.

I have now had a weeks work at the Metallurgical Laboritory & next Thursday my first lectures (Nat. Hist.) begin. I am glad they do for I find it very hard to work at anything in the evenings when I have no regular work cut out.

The Laboritory is decidedly hard work but I think I shall like it especially as it indulges to the full the taste which I always posessed for melting lead & other metals & working with fires. It is intensely practical, seeing that you have to light your own furnace of a morning (in which art the Lab. assistant gave us a most valuable & suggestive lesson) & stoke it all day, breaking up coke to a suitable size & shoveling it in as occasion may require There are only 12 places in the laboritory, & they are all taken up, & as usual many applications having to be refused. Several of us have provided ourselves with blue spectacles to preserve our eyes from the furnace glare, which is very great when engaged in assays which require some time & constant watching. The training will be very usefull, especially for commercial purposes as you can make more assays in a morning of many substances than you could chemical analyses in a weeks hard work.

I have of course returned to Mrs Guest & find her quite well. Mr White has not returned this session & so Mr. Phillips[148] besides myself is the only lodger This I think will be a decided improvement with regard to comfort.

I wish you would not make such a fuss, & talk so much about my little success last session, as it was not really anything so great & I hate to go under false pretences. By the way I have not heard anything more of Maggie Freer. I suppose that she must have returned by this time. I am sorry I missed her, & would have called had I known when she came back to town. Please tell her so if you see her.

If you have time I would like very much another pair of bedsocks such as you sent last winter. I found them exceedingly comfortable, & useful not only in bed but to put on over my slippers when the weather was cold.

I have not yet found time to call on any of my friends here but must try to do so soon especially Dr Bigsby & Dr Davis. I saw Sir C & Lady Lyell yesterday for a moment at the museum. They have just returned to town, & Sir C. is looking much better than when I Saw him last at Keswick.

[148]Arthur G. Phillips was the son of metallurgist J.A. Phillips and a fellow student at the Royal School of Mines.

I intend to send by this post my Cumberland sketch book for your inspection. As I think I told you I found no time for coloured sketches but made a few rough pencil ones which will serve me as mementoes of my Summer You ask where I am reading my "texts" at present. I have been rather changing about lately, but am quite open to suggestions on the point of reasonable.

I took a walk across to Battersea Park this afternoon & saw for the first time the sub-tropical garden, in which are collected together a number of plants of warm regions which will live outdoors in summer & are kept under glass during the winter. It was very pretty though a good many of the tenderer plants had been already removed.

I saw in the shrubbery of another part of the <gar> park a tree of "Pear hams" I recognised it at once from its different habit of growth <to>, & [supercosity] of fruit to the English thorn. The boys had found it out, & pulled all off that could be reached from the ground.

Yours with love

George Dawson to Anna Dawson, London, England, 22 October 1871.

My Dear Anna

Firstly can you tell me who the Mr Jeffreys is that mamma speaks of in her letter. For though he is mentioned twice I have not the remotest notion.

I was very glad to know that your lithographic venture has so far been successful, & wait anxiously to see your plates when they are printed. As you say the are fossils I suppose they are plates for Papa's devonian plant memoir.

I Called on Dr Bigsby yesterday morning & saw Himself & Mrs B. Both seemed pretty well, & Dr B working away as usual, How wonderfully dilligent he is considering his age. They enquired kindly after you all.

Last night I was at Sir C Lyells It turned out to be quite an evening party, dancing &c. It was Leonard Lyells 21st. birthday. Huxley,[149] Tyndall,[150] Dr Frankland &c. were there & so it was quite a distinguished throng. I always feel it a bother that I never learnt dancing, it is so uncomfortable to have to go dodging about on the outskirts & looking at books you dont' care to see.

I have been working at bullion assay at the Lab. last week. It is pretty clean work, & though rather troublesome on account of the numberless weighings I like it very much. It involves the using up of some money in samples, but the silver is of course recovered again at the end of the operations, & can be sold.

I will remember to write to Dora & Nina but at present have really nothing whatever to write about. It was very kind of Dora to write to me as she did.

I see by the papers that there has been some fenian row at Red River[151] As nothing more appears I suppose it has been quelled. At any rate I hope so as it might if serious cost much trouble & money.

You say you dont know what I intend doing with my scholarship. I thought you would have heard that Papa advises me to do nothing at present but keep it for eventualities.

Oct 26

[149]Thomas Henry Huxley (1825-1895) was an eminent British natural scientist who was also one of George's teachers at the Royal School of Mines. Huxley was a renowned scholar who wrote extensively on a wide range of subjects from palaeontology to philosophy, lectured widely, and received many awards and distinctions.

[150]John Tyndall (1820-1893) was another prominent British scientist whose major research was on the effects of solar and heat radiation on atmospheric gases. Tyndall was also a renowned popularizer of science, communicator, and noted alpinist.

[151]The Fenians were members of an Irish-American movement dedicated to Irish independence. In 1871 a faction led by "General" John O'Neill launched a raid into Manitoba but a United States detachment of troops from Fort Pembina followed and arrested O'Neill and dispersed his followers.

Please thank Papa for me for the copy of the Illustrated sent, & Mamma for Papas speech which I read with great interest.

With love to all

Your affectionate Brother

George Dawson to Anna Dawson, London, England, 3 December 1871.

My Dear Anna

I received the card with Miss Barbers address & thought it time last Saturday to try & find her out. Commercial St [...] was however rather wide, seeing that the street is about a mile long. I first by dint of enquiry found my way to [...] which is quite in the East end & not far from the docks. After walking up the street some way & keeping a look out for likely places, I saw in large letters on a building next to a church, "Home for Boys." Thinking myself lucky to find the place so easily, I pulled the bell. The door was very huge & heavy & had a great slit with *contributions* written up over it. After a time it was slowly opened by a *very* small boy, & is asked if this was the home of industry.

"No sir its further up the street" so on I went. At one place in a sort of vacant lot a whole fair of old women were assembled each with a great pile of dilapidated boots for sale. Generally gaping with holes, but all brushed up to look their best. At last I saw a policeman. I asked for the home of Industry. He was an irishman. "Industry — Industry –" he said, & then a sudden light breaking on him. "Oh! youl' mean the home of In-*dus*-try. The next door past these old women with the boots" I felt sure I had found the right place this time for the windows were covered half way up with paper bearing texts in blue, <about> the letters about a foot long. Inside it seemed all confusion, packing boxes, & dirt. But a door was opened & I was shown into a parlour. This also seemed in rather a confused state. The table was covered with articles of fancy work, which, (excuse the pun) I dont' think anyone would fancy. Impossible & gaudy baskets *etc. etc.* In one corner {of the floor} about three cart-loads of bibles & hymnbooks stacked away. An austere female with a promising moustache & a complicated erection in black crepe on her head appeared.

I asked for Miss B. & all I could gather from her rather incoherent account was that Miss B had not yet arrived there, but that some parcels were sent away the same morning & perhaps mine was among the number. Another austere female came in & corroberated the statement about the parcels. So in mortal fear lest I should be persuaded to buy some of the aforesaid fancy articles. Forgetting even to leave my card, {I bolted.}

No parcel has come as yet & so I suppose mine could not have been of the number sent. No doubt I will hear from Miss B when she arrives in person.

I also went yesterday to the Davies's at Regents park. Dr D was not at home & I was exceedingly surprised to hear of Mrs D's death. Which took place at Frome on the 21st. ult. She was down at her daughters, who is very ill, & was herself suddenly taken unwell & died in 24 hours & before Dr Davies could get down.

You will have heard of Mrs Bigsbys death. I only knew of it the other day. It took place about a fortnight ago.

I was at Col Lyells for dinner last friday evening. Pleasant enough, but giving me the pleasure of writing out my notes last night instead of getting some time for reading.

With best love

Yours affectionately

George Dawson to Anna Dawson, London, England, 28 December 1871.

My Dear Anna

Many thanks for your last letter, inclusive of the scolding. You see I get so tired of writing all about myself, & hate the sight of the numerous *I's* covering the paper. If I was accompanied even by a dog I think I should use *we*. The opera which you ask about was the

Huguenots[152] & was very pleasant to see, & listen to. Some of the best singers (hoping their names may be spelt approximately right) were Mlle Titiens.[153] Trabelli Batini & Senr Foli[154] & others, names forgotten. Trabelli Batini I think I liked best, she made a capital page Urbano. The singing was of course all in Italian but we had english translations to read from when we liked. It was rather absurd in one place just as the Huguenots & other soldiers were going to fight, to see a whole troup of Ballet girls rush in & seperate them, & then soldiers & all begin to dance.

On Monday Evg last I took dinner at Col Lyells. It was quite quiet only Sir C. & Lady Lyell besides myself. Sir C went home early on account of <the> his eyes which are hurt by much night light, & then we played various games till about half past ten. Lady Lyell kindly took me as far as the Portland Road Underground station, on her way home. She also presented me with a little callender with changing Cards. Mrs Lyell was so kind as to give me a very pretty little ink-bottle with closing top. Mrs L. asked me to come again next evening if I liked to a charade party, & also on New Years day. I got out of the charade party as it is rather a bore to go twice to the same place on consecutive evenings, & I meant to do some reading. Tueseday morning I began to read, but being boxing day all the shops were closed & the streets like Sunday, everybody going about in their best clothes. Mrs Guest also who is taking her Xmas out in a go at Bronchitus was coughing like a good one in her room. So under the combination of circumstances I started off, & went to the Crystal Palace about 2, o'clock. I spent the afternoon & had some tea there in the midst of a boiling multitude got back here by 8 o'c & did a good evenings work. There was a grand Pantomime but I did not go in for "reserved seats 6d" but had a look at the new aquarium which is well worth seeing. I afterwards saw part of the final transformation scene of the Pantomime by standing on top of one of the refreshment tables. There was a huge waterfall with any amount of real water coming down over a hill of about 1000 feet high among palms ferns &c. &c. They had a decorated bust of the Prince of Wales in front & played God bless the P of W on the organ as a finale, (encored of course). The library at Jermyn St is now open again. I was there yesterday & am going off as soon as I finish these letters. Please thank Eva very much for her illuminated Card.

With love your affect brother

George.

{I began with the intention of thanking you for your very pretty picture, but have gone on & on till I had almost forgotten. I like it very much, & though no judge, think It very well done. The only thing I see to criticize, & this will <probably> I hope be more flattering than unqualified admiration, is the sky. The clouds I fancy are a little massive. Your litho's are capital were they drawn on transfer paper?}

George Dawson to Anna Dawson, London, England, 24 March 1872.

My Dear Anna

We have been getting such delightful weather here. Thursday last it was very cold, & in the afternoon began to snow very heavily. About 2 inches came down & remained on the ground till half through Friday. Then yesterday it snowed again quite heavily at intervals through the whole day, but melted almost as it fell. It was quite doleful to see the young leaves on all the trees crusted over with snow. In the South of England many of the fruit trees are in bloom & so I

[152] An opera in five acts by Giacomo Meyerbeer to a libretto by Eugène Scribe and Emile Deschamps, produced in Paris in 1836, and set in Paris and Touraine in 1572.

[153] Therese Tietjens (1831-1877) was a German soprano who had a strong and flexible voice.

[154] Allan James Foli (1835-1899) was an Irish bass, who possessed a powerful voice of more than two octaves.

fear much fruit will be lost. Just fancy too yesterday being the boat race day. A good many people thought it would be postponed, but it took place notwithstanding, & I dont' know that any snow fell during the time they were actually rowing. Cambridge won again, for the third time now, but they have still a good many Oxonian victories to make up. I need not tell you, considering the weather that I did not go, in fact I had not any intention of going even if fine. Having seen it twice was I thought sufficient. However many undaunted people did go, as I found to my cost, for I happened to get into an underground train which was packed nearly to bursting. All the seats filled, & as many people as could find room standing up between.

I had a very kind note from Ward the other day, inviting me to spend my Easter Holidays with them at Keswick. They have he says just got permanently settled in a house a little way out of the town called Greta Bank Cottage, from which the three highest mountains in England can be seen. However I do not of course intend to go. Time being too short, & distance too great.

I received a present of shortbread from Edinburgh last week, made I believe by Maria Bell. It is very good, & I must write & thank her for her kindness. I heard by David that they had quite an accident at Gilmore Place some weeks ago. Some drain which carried away springs from under the house got stopped up & flooded the lower story with several inches of water. Nobody could find out where the damage lay, & for several days the house had to be kept dry by pumps. The "consequences were" a very bad cold all round.

I am rather sorry that it has been decided not to take a house at the sea-side, as I fear I may be so late as to get you possibly into very hot weather. You see The exams here will not be over till the last week in June sometime, & If I get a steamer conveniently I cannot be home till somewhere about the middle of July. However you will be best able to judge.

I am very glad to see that the mountain park is in so forward a stage, & hope it will soon be a fact. The Redpaths were among the timber cutting delinquents, as I can testify but they took care to do it out of sight.

With very best love
Your affectionate brother

George Dawson to Anna Dawson, London, England, 19 May 1872.

My Dear Anna

Last weeks letters only arrived yesterday (Saturday) evening which is exceptionally late for the season. As I expected the letter which was promised by intermediate mail came also by the Canadian marked "too late". I also received a copy of the Canadian Naturalist & a Gazette[155] with an account of the convocation proceedings. I like your picture of the Post Pliocene shells very much. It must have been a great deal of trouble to draw them so accurately "on stone".

I have just come in from taking a walk all round Battersea Park. On the way I have seen no less than five distinct street preachings going on! The first man was a temperence lecturer & he had a pretty large audience. He was relating his experiences &c. & telling funny stories.

Next there was great discussion going on between a man who calls himself a "Humanitarian" & lectures every Sunday near Chelsea Bridge; & another man of more orthodox opinions. Then there was a man in a little pulpit with some texts in very large letters painted in front, discoursing away at the very top of his voice to a very small & scattered audience. I have seen him often before. He is always in a great persperation & without a hat, has hardly any front teeth. I conjecture they may have been blown away in moments of special vehemence. Going over the bridge I came on another preacher. He was talking in a more rational way & had a fair number of people listening. Lastly another man at one of the gates of Battersea Park preaching to a very select audience in a feeble voice. I had quite a long walk altogether & finally got into a region of desolate fields waiting to be built over. Far drearier than the deasert of Sahara could by any possibility be.

The Mining examination comes off next Saturday. I shall be glad when it is over for I really dont' see how the Natural History & Applied mechanics are to be got into examination order before the exams in those subjects take place. The Mining seems a most unsatisfactory subject, so little you can grasp, or get hold of in any way. So large & so intangible. It is a great disadvantage to me that I dont know anything about it in a practical way as many of the men do.

With love to yourself & all at Home

Your affectionate Brother

George.

P.S. Please remember me to Nina I feel very much ashamed of not having written before this.

George Dawson to Anna Dawson, Middlesborough, England, 11 July 1872.

My Dear Anna

Here I am at last after a week of awfully hard work in London. I had no idea what a business packing up all my goods would be. Every day something new turned up. The weather all the time was as hot as it ever is in England & I had so much business in the city & all over London that I was kept running about nearly all day & had only the evening for actual packing.

Having a day or two to spare before the 16th when I sail from Liverpool by the Caspian, I decided to spend the time somewhere & to see what I could of mining &c. Cornwall was rather out of the way, & so I chose to come here. This place is a huge workshop. It has grown up within the last few years & now has some 40,000 inhabitants. Here all the iron ore from the Cleavland district meets the coal from Durham N. Castle &c., & there are several hundred blast furnaces here & in the immediate neighbourhood besides all kinds & descriptions of rolling mills & iron manufacturing generally. At night the sky is quite lurid with the glare, & it looks as if great conflagrations were taking place in all directions.

Young Huxley[156] after finishing some business connected with his future situation at Sheffield has met me here & we are going round together & seeing what is to be seen.

I should have begun by telling you the results of the exams, but really it seems such a long time ago now that I forgot to do so. I have got through all right in both the Mining & Geological divisions. I have got the Ed. Forbes medal & prise of books in Nat. Hist. & Palaeontology, & would have had the Medal in Mining but there is some regulation about not allowing a student to have more than one. I dont know how I managed to get so high in Mining. In Nat. Hist. there were five men went in for the exam. I was the only one eligable for the medal, as the others had not done palaeontology. However I managed to head the list & Prof Huxley told me afterwards that I had done very well & was considerably better than any of the others. Leonard Lyell was next below.

I got a note from Nina last Monday morning. It had been Sent to a wrong address & had been detained a day or two, finally arriving with "Not known at 26 Halsey St written on it." It seems through some mistake you had given Nina the wrong number. Monday was my last available day in town so I went round in the afternoon. I got out at Notting Hill gate & after a good long walk found the place. It appears I should have got out at Notting Hill on the Hammersmith line, but I did not know there was a station of that name. I was so fortunate as to find Nina in. She was looking very well & hardly at all changed since I saw her 3 years ago in Montreal. Alfred was down at Wimbledon in Camp. I saw your photo, & like it *very much*. It seems very well taken, & a great success every-way. I am sorry I did not get Nina's note soon-

[155]A scientific magazine, and a Montreal newspaper.

[156]James Huxley, another student at the Royal School of Mines, was the son of Professor T.H. Huxley.

er as I might have been of some use showing them various things & places. Still as my time for the last week was so very fully occupied perhaps it <was> {is} just as well.

I must cut my epistle short as it is now nearly breakfast time & we have promised to be ready at 9 o'c. to go under the guidance of a fellow called Charlton & see some of the furnaces &c. Charlton is a fellow student from the School of Mines. He gets the mining medal. <We> & has passed very well in his other subjects. We met him quite by chance in the train last night as we were returning from a visit to the Upper Leatham iron mine. He lives here & knows all about the place.

With love

Your affectionate brother

George

P.S. I suppose you will hear of me next at Halifax. If I find a steamer convenient I may write again from L. pool before sailing.

GMD.

Officers of the British Boundary Commission, Dufferin, Manitoba, Spring 1873. George Dawson is third from right, top row.

Scouts of the British Boundary Commission, Dufferin, Manitoba, Spring 1873.

Indian Wigwams on Island, in Lake of the Woods.

Maxim Marion, Métis Guide.

Trader McPherson's Family, Northwest Angle, Lake of the Woods.

Ice Cutting Party, British Boundary Commission, Red River at Dufferin, Manitoba, 1873.

Red River Transportation Company's Steamer Dakota *at Dufferin, Manitoba, Spring 1873.*

Chippewa Indians at Dufferin, Manitoba, Spring 1873.

Camp at Turtle Mountain Depot, Manitoba,
British Boundary Commission, 1873.

Observatory Tent, Zenith Telescope, 1873.

Sappers building Boundary Mound, 1873.

Métis Traders, 1874.

Log Hut at Métis Settlement, Wood Mountain, Saskatchewan, 1874.

Depot on East Fork of Milk River, Saskatchewan, 1874.

Dead Crow Indians on Boundary Line, 640 Miles West of Red River, July 31, 1874.

Grave of Sioux Indian, 1874.

Assiniboine Indian Camp, 1874.

GEOLOGIST/NATURALIST ON THE BRITISH NORTH AMERICAN BOUNDARY COMMISSION

George Dawson's first important scientific appointment in Canada was as geologist and naturalist to the British North American Boundary Commission, which marked out the international boundary between the United States and Canada in 1873 and 1874. Dawson devoted his efforts especially to the geology of the region traversed by the commission, a large part of which was then unknown. He presented an admirable account of his work in his report entitled "The 49th. Parallel,"[157] *which has since been called a classic. Along with his geographical and geological work, Dawson made extensive studies of bird life, sending many hitherto unknown specimens to the British Museum.*

The following, some of George's letters to Anna written during his commission fieldwork, provide interesting descriptions of his varied experiences, including lively illustrations of travel in the field.

George Dawson to Anna Dawson, Dufferin, Manitoba,[158] **31 August 1873.**

My Dear Anna

On my arrival here Thursday last I received such a budget of letters that I am at a loss to know to whom I owe the first reply.

First to answer your questions. The mosquito oil is so far good that if you smear yourself with it no mosquito will light for about half an hour. At the end of this time it looses its effect & you have to repeat the dose from a small bottle carried in the pocket. The remedy at best is worse than the disease & its frequent application destructive of comfort & clothes.

I feel quite well or rather better & when I read your groanfull letters began to think I had better not alarm you by any more accounts of my goings on. So far I have never ceased to enjoy myself.

N.B. I am going to look over your letters before closing & if I find any more questions will answer in a P.S.

When I wrote last I was just leaving the N.W. Angle for Rat Portage.[159] I took only one canoe with Begg the half, or rather 1/8th breed who had been round the lake with me before, & an indian who unfortunately could talk hardly any English. The trip took 7 days & nothing very remarkable occurred. The first few days were beautifully fine & warm, then Came four days of broken weather with wind & rain during which time we were dodging along behind islands & lee coasts to make any progress at all. One evening after a heavy afternoon

[157]*Report on the Geology and Resources of the Region in the Vicinity of the Forty-Ninth Parallel* (Montreal: Dawson Brothers, 1875).

[158]A Canadian post on the Red River, north of the international boundary, which served as a base of operations and winter quarters for the British Commission.

[159]Kenora, Ontario.

of rain camped in a dripping wood with spongy moss & rotten leaves saturated with water which is not the essence of Comfort. An indian appeared in his canoe & on being asked for fish said he had some dry sturgeon. He set off for his camp & returned in about half an hour with the fish, for which we gave him some flour, & also let him have cup of tea &c. The next morning I awakened by 5 to hear a great jabbering & found my indian conversing with three others, including our friend of the night before, & no doubt induced to come by his good report. One of them had some fresh sturgeon for which he got some pork & the rest had what remained in the teapot &c when breakfast was over. A day or two after my return to the Angle I recognised the man from whom we had got the dried sturgeon & found that he had come in in great alarm to report the breaking out of some infectious disease among the indians on the lake, & as it turned out, at the camp near which we had been, & from which our indian visitors had come. A woman, he said, had taken it first & broken out in spots & died. This was a day or two before our visit, & thinking the disease infectious they had burned her body completely up. A few days afterward the womans three children took the same disease & broke out over the head & legs. This had alarmed the indians who thought it must be small-pox & came in to report. Very probably however it may only be measels which is a very fatal disease among them. I have not heard since how it has turned out.

I left the Angle finally to come here on the 18th. taking only a single canoe with Begg & Spearman for Crew, having sent Duckworth[160] back here by road in charge of heavier baggage belonging to my party & some to Mr East's[161] which is going up White Mouth R.

Hearing that the Tug was going out for Hungry Hall the same morning I made arrangements for a lift & got the canoe & stuff put into a boat in tow & carried down to Flag Isd which is at the mouth of the Angle Inlet[162] & quite half a days paddling from it. (By the way, I must put in a parentheses to tell you that quite a grand side wheel steamer like one of the market boats at Montreal is now almost ready to ply on the Lake of Woods. She has been building several years at Fort Francis & arrived on her first trip to the angle a day or two before I left)

Well, leaving Flag Island we coasted south & as the day was very fine made a long journey & camped after dark just inside Buffalo Bay.

The next morning was broken & windy & with our canoe heavily laden we struggled across the bay to the mouth of the Reed R which we reached just as it was becoming too rough to go further. That river turned & twisted in a remarkable way & had also a pretty strong current, & altogether we had a pretty good days paddling before reaching a part where it was only about a canoe length wide, where we camped.

The next day though we started early it took till about 9 o'clock to reach the source. The river got smaller & smaller & would loose itself for a time among the rushes of the swamp. I had anticipated some difficulty in finding the beginning of the Portage but just as we came to where the river ended there was a sort of track through the reeds by which canoes had evidently gone. Entering here it soon became too shallow for the canoe to run & so knowing it was the commencement of the portage all jumped out & began to drag the canoe along. There was a regular little rut where canoes had been dragged before & water running down it like a brook.

Soon however the swamp became too shallow for the canoe & it was necessary to take half the stuff out & carry it a few hundred yards. Then the swamp became deeper again & all the stuff was put in once more. The hard bottomed shallow piece formed as it were the edge of the swamp basin. Then began the tug of war. Spearman & Begg tackled themselves by ropes to the thwarts of the canoe & pulled manfully.

[160]Sapper Duckworth was a taxidermist for the commission.

[161]D'Arcy East, formerly an officer with the Royal Artillery, joined the commission during the 1873 season for a special survey of the Lake of the Woods.

[162]Northwest Angle Inlet.

I pushed behind whenever it came to a tough place. I had a pair of long beef mocasins but I soon found they would not do as at each step I had to lift about 1 imp Qt of water, so taking them off I tied a pair of stockings round the ankles with string & in this rig got along capitally. This was a regular muskeg covered with wiry grass & moss with small groves of tamerac here & there. In general I should say the water was about knee deep, but every now & then you got into a waist deep spot. In some places the bottom moved for about 10 ft all round when stepped on, & often in these places a patch would appear like dry ground but would sink under your weight giving out copeous streams of sulphuretted hydrogen. In some spots for a few paces there appeared to be bottomless swamp muck, & then one had to rest most of weight on the canoe & haul over. There were fish too poking about among the grass, but I could not catch any they seemed to be young pike.

About 12.30 it was considered dinner time, so choosing a tussock of grass raised above the rest we made a fire of dead tamerac sticks & boiled the kettle for tea standing knee deep in water all the while. Then we went on again & on & on till I thought the swamp was never going to ease. It was like walking through very deep snow but additionally uncomfortable. At last we got out on a great open grassy swamp & saw the main woods on the W. side at what appeared a great distance However by the aid of numerous rests, we got over it at last, & just as dusk found a little rivulet rushing out of the swamp *Westward!* This was the beginning of the Roseau R.

It got dark before a good camp could be found & so we were obliged to sleep in a poplar grove where the soil was rather squashy, & where there was no place for a tent. The night was however fortunately fine. They call the Muskeg Portage 8 miles long but I am mistaken if it is not much nearer 10. At any rate it took us from 9 A.m. to 6 P.m. to "do" it.

I had meant to give you a description of the rest of this trip, which was rather more eventful than usual but find it is getting very late & must close with a short summary.

Next day occupied cutting windfalls across the river breaking up beaver dams &c. & untangling jams of trees which the beaver had cut down & which nearly stopped the R in places. Had to make two portages past impassible jams. Next day got past Roseau L. & out into a great treeless swamp where had to carry wood in canoe to cook. Slept on a mud bank in same where no place to put up tent. Rain & wind before morning. Got thorough ducking. Crept under Canoe. To stormy to launch canoe till 10 A.M. Then got off & by 2 P.M. get to place where plenty wood. Made a big fire & got things dry. Millions of ducks &c. Next day got into rapids small, but next two days occupied in getting the canoe down about 20 miles almost continuous rapid full of boulders. Wading in water most of time letting the canoe down along the shore & then running such pieces as were practicable & had calm water at the bottom. Next day paddled about 28 miles & got on our way about 9 miles. Next day in Rankine letter.

Your loving brother

George Dawson to Anna Dawson, Dufferin, Manitoba, 5 November 1873.

My Dear Anna

Letters from home received lately have all began by excuses for not writing before. I think it might be right in me to begin in the same way as it seems to me very long since I wrote last.

You will see by the superscription that I have got back here to head quarters, having arrived just a week ago today. None to soon for comfort as we are now experiencing quite winter weather & I fear too late for any indian summer break.

I have been engaged here getting things put in order & packed up & had hoped before this to be on my way down to Montreal. Packing however is rather an arduous operation, & besides there is really great difficulty in getting away from here at present at all. The winter has come on quite unexpectedly early, & communication by water stopped by the freezing in of the boats on the river. The tri-weekly stage is the only remaining chance, & so many are seeking an exit by its means from Garry[163] & points to the North that it is almost impossible to get a place on it at all. Another trouble is that as I shall have to bring along several heavy boxes, the expense by stage would be absurdly great. However I hope to surmount these difficulties within a day or two in one way or another.

The quarters here have grown into quite a little town now by the erection of large stables, & stores & all is bustle & preparation for the winter.

I was much surprised by the advent of Mr Selwyn[164] the other day. He was on his way to Montreal from the Saskatchewan region, but having been frozen up in the Red R. Hired carts from some half-breed & gon on this far. He was here a couple of days before he could make arrangements to get on & I heard a good deal about his explorations on the Saskatchewan &c. I should have liked very much to have gone on with him & had I had a day or two's warning <of> might have done so. I have got so far behindhand in my chronicle of events that I cannot now begin to make up lee-way but must trust for that to oral communication. We had a pretty rough time of it for the last week on the way in, the ground being covered with snow several inches deep & the weather very cold with almost constant snowstorms & high winds. Nearly all the prairie along the line is burned & for a picture of desolation I dont think burnt prairie partially covered with snow can well be surpassed. There was of course no grass for the horses & what with little hay & very little oats, many of them "played out" & several carts & animals were left behind.

When camped in one place with plenty wood close at hand & a stove in the tent one may be moderating comfortable even in such weather, but on the move every day & all day long, with neither of these advantages & often obliged to camp after dark, things become mixed. Bread & beef & ink & everything freezable being frozen is not conducive to comfort, & getting up in the morning before daybreak with the thermometer far below the freezing point is unpleasant. One night the temperature fell well below zero, & since our return here, the minamum has once been -18. However all went well & the whole of the men are now back here without accident of any kind.

Parties coming in would have given good opportunity to some caricaturist of motly groups. All sorts of make shifts & rig-ups being the order of the day. Some men with torn clothes, others with gaping boots, some with blanket mits & some with none at all. Mackintoshes below coats, & head pieces that had seen a summers wear. Stopping for lunch every man with a chunk of bread thawing it slice by slice at the fire.

You say my letters meerely relate to facts & dont say anything as to what I am thinking about &c. I suppose this is a true bill, but really I think travelling about over these plains does not conduce to much thought. Ones chief idea is to get warm & something to eat, which having been accomplished you feel sleepy & so wrap yourself up in your blankets & go to sleep. Subjects for reflection being few but often repeated become monotonous. Having once worked out a train of thought anent buffalo skull, a burnt prairie, an unburnt prairie, a tuft of grass, a prairie chicken;[165] one has just to begin & go over the list again & again, & therefore one has either to ponder on some abstract question not at all connected with the place, or rest content to go along without thinking at all.

Please tell William that I shall be very glad to subscribe to the University Gazette to be sent to my address here. The specimen copy sent by him never reached me. Having probably lost its address or been mislaid in the post office somewhere.

With love to all & hoping soon to follow

Your affectionate brother

[163]The Hudson's Bay Company's Upper Fort Garry, part of present-day Winnipeg, Manitoba.

[164]Alfred Richard Cecil Selwyn (1824-1902) was director of the Geological Survey of Canada from 1869 to 1895, when Dawson assumed the position on Selwyn's retirement. Under Selwyn's direction, Dawson and his contemporaries began the immense task of surveying the western portion of British North America.

[165]Greater prairie-chicken, *Tympanuchus cupido* (L.).

George Dawson to Anna Dawson, Dufferin, Manitoba, 19 April 1874.

My Dear Anna

It seems a long long time since I left home & I believe I have never yet written to you. At present one would certainly need to follow the plan of having a letter always ready, for the bad state of the roads has so disorganized the coaching that one never knows which is mail day.

Well to begin with, we had a very pleasant Journey all the way from Montreal to Morehead.[166] The trains being on time nearly all though. *We had one day* from 8.30 A.M. to 10 P.M. *to spend in Chicago* & walked about the streets & looked *at things* till we were tired. The streets and shops are very handsome, but all the libraries museums &c. were burned up during "the fire" & as the parks were not green, there was nothing in particular to see. The place is built on low flat ground & has a foggy, smoky, heavy sort of atmosphere from its proximity to the lake. Morehead was reached on Saturday evening & we had thought to spend Sunday there as the regular time for the stage to start is Monday Morning. The roads being bad however *the stage went out at 2 o'c on Sunday afternoon* bound for *Georgetown* the first station on the way North & 16 miles from Morehead. The road was quite impossible so we took to the prairie sod & drove miles & miles round the ends of flooded coulees & through half frozen swamps. The driver had not been on the road for two years & so did not know the position of the new bridge at Buffalo Creek[167] 3/4 mile S. of <Morehead> Georgetown. As it began to grow dark he began to be doubtfull & finally steered toward a light on the river edge & enquired. After groping about some time we found the place with a big half thawed drift accumulated in the valley of the stream. Going down <to this> through this to the bridge the coach got completely stuck, so we walked on to Georgetown through the most frightful mud & stayed there all night. You cross the Red R at Georgetown & go on to Garry on the west bank. As the ice was too rotten to take the stage over, in the morning, we had to get across to another stage on the other side, & this was accomplished partly in a little boat & partly by walking on the ice. The new stage was then found to be a very old one & took the driver about two hours to tinker it up. At last we got away at 11 o'c & got on to Elm R. the next station. Here we had expected to get dinner but found none ready & so were obliged to go on to the Next stage Goose R. which was reached about 7 P.M. & here we made supper & dinner combined.

Breakfast dinner & supper are much the same except in name on this line. You find generally fried pork remarkable fat, potatoes, & if you are remarkably lucky eggs. You may always get bread & generally dough-nuts. The whole spread has a greasy appearance, & the remains of the eggs of two or three seasons may be detected on close examination sticking between the iron prongs of the forks. The tea so far as my experience goes is either absolutely tasteless, or as bitter as gall.

We left Goose R at 7.30 & drove on against a gale of wind from the north & a storm of sleet & snow. When it got quite dark it was impossible to keep {near} the road & we very soon lost it & wandered about on the prairie. The driver then tried to light the lanterns but found one utterly smashed, & the other with one pane out & not wind-tight. After several attempts we managed to stop the broken side with canvas & light the candle inside the stage. At last we came to a coule nearly full of snow & the driver judging that he must have passed the station & seeing a light behind, steered toward it & brought us at last all right to Frog Point.

The next morning we started at 8 A.M. the morning cold & a slight coating of snow on the ground. Three miles out the fore stage got into a water-hole in a coulé & the fore wheels pulling out, the driver

[166]Moorhead, Minnesota.

[167]Buffalo River.

flew off the box after the horses. The stage had to be unloaded entirely & after a good deal of lifting & prying we got the wheels put right again & went on. The next coulé was filled with a great snow-drift so the stage had to be unloaded again & drawn over the snow by hand on blankets & buffalo robes spread out for the purpose. Soon we came to Coule number three which was full of water with rotten ice over it. The horses were lead across & then the attempt was made to haul the stage over by hand, but the wheels cut through & the water coming upon the ice rendered it necessary to take off all the baggage again. Before we could break away the ice & haul the stage out two or three hours were spent. Arrived at Grand Forks at 5 P.M. & had dinner. Started out about 6 with a good team but the driver had been imbibing rather freely. We got on through several bad swamps pretty well but at last got into a very liquid bog in which the wheels on one side went down to the hubs, & the coach was on the point of turning over when we all scrabbled out up to our ankles in water. At last the stage got out & soon a light became discernable in the distance. The driver was now however hopelessly drunk & could not see the light. At last we fortunately got on the road & then the driver whipped up the horses & we rushed along fortunately without accident till the station was reached. Here having had supper we availed ourselves of the accomodation of the place & lay down on the floor in blankets for the night. The driver excused himself in the morning by saying that he did not "respect the man who would not get drunk sometimes on such a road." We made an early start & got along pretty well crossing the Big & Little Salt rivers[168] on the bridges which were just on a level with the water but not covered. One bad hole was met with during the day the fore horses fell <in> & then the <hind> others on them & the stage nearly on top of all. However no serious injury accrued & we reached Pembina about 9 P.M. & got on here before midnight.

Since my arrival I have been very busy with one thing or another. Captain Cameron[169] had several things to talk about & I am engaged in answering some questions on the country between here & the Lake of the Woods.

I do not think it likely that we will move from here before the fifteenth of next month. Our letters &c. will probably go via the Missouri, or by Union Pacific RR & Stage, to Fort Benton Montana. This fort is only 80 miles S. of the line & it is supposed that at least occasional communication will be kept up with it.

I hope you are much better by this time & that you will soon be going out which will be a pleasant change after having been in the house so long.

With Love to All

Your affectionate brother

George Dawson to Anna Dawson, Dufferin, Manitoba, 28 April 1874.

My Dear Anna

I have written at some length to William & so must content myself with meerely a short note to you. I think I have already told all the Knews.

I have Duckworth hard at work again shooting & skinning birds & hope to get a good collection together. Spearman my old servant has left the commission & I have another man called Nillson. I think he will do very well but may be better able to tell after a time.

I had a letter from Russel[170] yesterday. He has been surveying in

[168]Forest River.

[169]Donald Roderick Cameron (1834-1921), of the Royal Artillery, was British commissioner on the boundary commission. He was later commandant of the Royal Military College in Kingston, Ontario, from 1888 to 1896, before returning to his native Scotland where he died.

[170]Alexander Lord Russell (1842-1922) held a principal surveying position on the British Commission.

the Lake of the Woods region, & writes that he has made a collection of rocks &c., & wishes to know when I am coming up to Garry. I had no intention of doing so at all but may change my mind when the boats are running regularly, especially as the part of the country he has been working in is one I want information on badly for my map.

Life here would be sufficiently monotonous if one was idle but I have still a day or two's work at my reports & maps, & then have other things on the *tapis*.[171] Breakfast is generally about 9 o'c. Then those who have anything to do go off & do it & those who have nothing set round the stove in the upper hall, (our drawing-room) & smoke. Lunch about 1 or 2 as the case may be. About 4. go for a walk, or to shoot, or to look at the river or any other excitement. Dinner generally about 7 P.M. Then adjurn to the *Drawing room* again & cards & chess, & talk & reading & whisk toddy & smoking occupy the remainder of the evening.

Spite of the Commissioner's[172] most strenuous efforts we have got our outfit allowance. It amounts properly to $500 but he has managed it so that we have to pay for Horse tent & saddle out of it so that some is absorbed. We have also the pleasure under a recent decision of somebody of paying income tax out of the half of our Saleries supposed to derived from England. However it amounts to a bagatelle. A lot of furs were got from the indians at Turtle Mt Last winter by Hill who was trading there for the Commission. They consist chiefly of Muskrat, but there are a good many Mink & some badger, bear, deer, wolf fox &c. Several of us went over to the store & picked out a few the other day. Herchmer (commissary)[173] valued the mink at about $2.50 per skin, <&> but the commissioner a day or two afterwards became dissatisfied & gave out that they were to be charged $3.75 all round which is the Montreal market price for the best quality of prime skins. Accordingly we returned them all to store, & I only hope he may get that much for them somewhere else. I have got a first rate skin shirt & am getting a pair of trousers made to match. On looking over my clothes I find I have now accumulated such a *heap* that it would nearly take a special wagon to carry them all out on the line.

Hoping to hear soon & favourably of your health & hair.

Your loving brother

George Dawson to Anna Dawson, Souris River, Manitoba, 5 June 1874.

My Dear Anna

Here we are in the leafy month of June almost without knowing it so fast does time run on. We hope to get across the first crossing of the Souris tomorrow & at the second crossing which is about 50 miles west from here anticipate no difficulty in fording the stream. Then about fifteen days should take us to Woody Mt which is rather past the end of last years work

Capt Anderson[174] got a copy of <an> a U.S. officers track survey made last year, before leaving Dufferin. It runs northward across the Northern part of Montana As far as the line through an unknown country. The officer says in his notes that several times they had to sit up all night fighting the buffalo to keep them from going into the pools <for> {on} which the camp depended for water. The buffalo had

[171]Under discussion.

[172]Presumably Donald Cameron, the British commissioner. See p. 106.

[173]Lawrence William Herchmer (1840-1915) later attained the position of commissioner of the Royal North West Mounted Police, which he held from 1886 until retirement in 1900.

[174]Samuel Anderson (1839-1881) served as chief astonomer to the commission.

been hunted by indians. It would seem from this that we should not want for fresh meat, but that the country must be remarkably dry.

I crossed the river yesterday with Dr Burgess[175] & two men, using the extemporary boat fitted up for bridge making which consists of a wagon body covered with a tarpaulin, & investigated one of the old Mounds known to the half-breeds as Mandan houses.[176] We found portions of three skeletons. One much broken up. Another apparently that of a woman or very small man, nearly perfect, the third a very young child, fragments. The most perfect skeleton was placed in a sort of sitting attitude, & the three seem to have been covered by a sort of roof of sticks & posts the wood of which in a rotten state may still be seen. The appearance did not seem to point to a greater age than say 50 years, but the mode of burial is quite different from that of the Sioux who inhabit this part of the country at present. The only things we could find with the bodies were, 1. a flat smothed rounded piece of bone like a paper cutter. 2. A large unio shell[177] from the river. The mound was over 50 feet wide & 6 feet high & must have taken some time to pile up.

Our diet is at present chiefly *bread* & a sort of imperfectly smoked *bacon* with some *cheese* & other condiments. I enclose a copy of the entire list. When we get out west some of the oxen will be sacrifised & replace pork by fresh beef. I cannot write more or I will exhaust my whole summers stock of paper.

Your loving brother

George Dawson to Anna Dawson, Wood Mountain, North West Territories, (now Saskatchewan), 24 June 1874.

My Dear Anna

It is difficult to sit down to write letters when one has not heard from home for so long & also when one knows that so long in time must be occupied before the letter written reaching its destination. However I hope that within a few weeks we may open up some communication with Ft Benton on the Missouri & get hold of whatever has found its way so far. This goes back by the old road to Dufferin along the line & will be a long time *en route*.

We are halted here for a day or two & very pleasant it is for a change not to have to get up quite so early or travel so long. I have nothing in the way of geology to interest me here but have found a few plants &c. & as William will be delighted to hear discovered a rather good spot for land snails.

Our usual order of march is something like this. Up about 5, breakfast on fried bacon bread & tea. Start about 6. Travel four hours. Halt till about 2 P.M. or during the heat of the day & during this time amuse yourself by lying under the shade of a wagon fighting the mosquitoes if hot weather, or lying in the sun in the lee of a wagon if cold. Reading if you have a book or very likely sleeping. Lunch on bread & cheese, or bacon, & tea with perhaps some stewed dried apples. Start again & travel three hours. Camp. Get things arranged in tent & pick out the piece of ground which looks the best shape for sleeping on.

[175]Thomas Joseph Workman Burgess (1849-1926) was first medical officer to the commission. Trained at the University of Toronto, Burgess later became a lecturer at McGill and medical superintendent of the Montreal Protestant Hospital for the Insane.

[176]In his journal Dawson further noted of these Indian houses: "A circle say 20 feet in diameter is dug out to 3 feet deep. Then sticks placed together as for a wigwam & sods & earth piled outside. The fireplace in the centre of the floor. The houses are made large as it is customary to bring the horses into one half of the building at night to prevent them from being stolen." (Allen R. Turner, "Surveying the International Boundary: The Journal of George M. Dawson, 1873," *Saskatchewan History* 21 [Winter 1968], 23).

[177]A river mussel of the family Unionidae.

Then supper or dinner as you may choose to call it on fried bacon &c. as before.

Of course the absence of suitable stopping places &c. modifies the perfect scheme a good deal, & one may also have an occasional duck to add to the bill of fare.

My trip from Wood End[178] was not a very pleasant one in some ways. I began well with the toothache which going away left a swolen face which rapidly increased till one eye was nearly bunged up & it was so difficult to chew that I had to subsist almost entirely on soft food. However it came to a head at last as a good sized gum boil & then happily passed away, leaving however still a little swel [page torn] which gradually diminishes.

We last heard of Rowe[179] several days ago when a man sent on with a light-rig for provisions caught us up. The injury appears to be rather a serious one & Dr Burgess fears to move him on for some time in case of inflamation affecting the brain. The hot weather has also been against him.

I should like to write to others at home & to William in Cape Breton but have really no more news. All goes so very quietly. Whenever another chance offers I will write again & then it will not be your turn. I suppose when you get this you will be at Kamouraska & I hope to hear how you like it, & that you have spent a very pleasant summer there.

Love to all at home & best remembering to friends in general.

Your loving brother

George Dawson to Anna Dawson, Three Buttes, Montana, 31 July 1874.

My Dear Anna

After so long an interval during which so much has happened that one might *talk* about, it is difficult to know what to *write*. Here we are at any rate camped at the Western end of the West Butte of the Three, with the Rocky Mountains in sight whenever you choose to climb up one of the minor peaks & see them. The Rockies are about 115 miles distant but on a clear day are beautifully defined & show great white patches either of snow or some light coloured rock. I move on again tomorrow, at least I intend to if the *oats* come up. They are close at hand now, having come through from Helena in Montana in charge of a certain contractor called I think Conrad, who arrived here last night.

Ashe[180] who is doing the survey work, now that Lieut Rowe is out of [count], came in last night with the report that in running a survey line he had come across twenty one dead indians who had evidently been killed some time. Conrad who is well up in indian news south of the line supplied the information that they were probably a party of Crows who had left the country near Ft Benton last autumn on a horse stealing expedition, had been surprised by an overwhelming number of Peagens (a sub tribe of the Blackfeet) & had never returned.

It was decided to send the photographers out to the scene & having in common with some others a spare day, "along of the oats," I thought I could not do better than go too, & have now not long returned from the trip. The spot was about ten miles from here on the open undulating prairie. The remains were in quite a mummified

[178]A supply depot approximately 260 miles west of Red River which later was a site of a North West Mounted Police post.

[179]Valentine Francis Rowe (1841-1920), an officer with the Royal Engineers, had joined the boundary commission as an assistant astronomer in the spring of 1873 to take charge of a special survey of the Lake of the Woods.

[180]W.A. Ashe was a well-known dominion land surveyor who headed up Rowe's topographic party during the 1874 field season.

state, the skin being stretched tightly round the bones, & the latter only protruding where the wolves &c. had been depredating, & where the scalps had been removed from the forehead to the back of the neck. The story of the fight was pretty plain, the crows, if such they were had evidently been on foot & their adverseries mounted.

The crows on finding themselves surrounded had chosen a sloping hill &, probably choosing badger holes already dug, had enlarged them & piled stones round the edge so as to form shallow rifle-pits. Their enemies had no doubt ridden round & round at full speed firing at them as they rode in indian style & finally succeeded in killing the whole party. The Crows had shot one of the horses, & probably some of the men but these had no doubt been removed. The skulls were all destroyed by being broken in & the bodies had evidently been cut & slashed in all directions after death. We picked up several bullets & iron arrow points, & fragments of shafts. The chamber of a revolver an old knife & some beads.

These Buttes & the line of the Milk R. bound a sort of neutral ground between four or five tribes & the indians never pass except in war parties. This accounts for their absence at present & for the fact that no recent camp fires or lately killed buffalo are seen in this part of the country.

I must distribute my favours & so please excuse brevity & believe me your loving

George Dawson to Anna Dawson, West Butte, Montana, 3 September 1874.

My Dear Anna

Here we are on the way East again having finished all the field work. We had our last view of the Rocky Mts yesterday, then 115 miles distant & just showing as faint blue peaks above the horizon.

There is quite a large camp here at present nearly all the U.S. parties being congregated previous to going home by the Missouri. They hope to reach Benton in 3 days & will then be 20 days going down the Missouri to Bismark in Mackinaws, or flat boats.

Of our side, the two astronomical parties went east yesterday, having work to do in erecting mounds & also in depositing "soup plates" 10 feet E. of each mound. We, that is to say self, Capt A. & party. Ashe & party & East with the scouts constitute the rear guard & are lying over here for a day to get some repairs &c. effected & pack up the remains of the Depôt. I say the remains for the indians have got most of what was left here. Only 3 men were left to guard the place & when some 400 Peagins came & camped down beside them they did not feel particularly strong. The indians did not actually take anything by force but came in & examined the things & said they must have so & so & they got it. In this way the whole of the tea & sugar was disposed of & most of the matches & a good many other things besides. The chief presented the depotman with various whips, lariats, &c. & the whole thing was perfectly amicable but much to the disadvantage of the commission. Fortunately we are ahead of our time & consequently have plenty & if the depôts at East Fork & Woody Mt are safe will be all right.

There are a good many buffalo here now but only in small scattered bands. One was killed yesterday, an old bull, & pretty thin, but not so bad as some beef I have tasted.

We have a regular marching routine now, & the waggons & carts go off in a certain order & at night draw up in a semicircular Kraal on the border of some coulé or stream. The tents are then pitched outside & the animals all tied up to the waggons at dark. The camp is called by the watchman at 4 A.m. when it is barely light, & always remarkably chilly. Breakfast before 5 & all packed up & off punctually at 6. March till about 10 or 11 according to distance necessary to make for water & feed. Halt for two hours & then travel on again for two or three hours. Making every day something over 20 miles.

If all goes well we will reach Dufferin on or after the tenth of October, much more probably after than on.

I have got a *few* sketches among the Mts. but very few. One seemed always to be doing something or going somewhere from daylight to dark & when you did take out your pencil it was only to jot down some view that *happened* to be seen from some camp or stopping place, & not one chosen as characteristic or remarkable.

I have become the happy? posessor of an indian pony, bought from the traders for $25. It *may* be a very nice beast when it learns a thing or two, but any way will answer my purpose in enabling me to ride or drive as I may please on the way in, & make it easy to leave the road if occasion requires.

We have had no letters for some time, but hear that the U.S. people have sent a waggon down to Benton to bring up *all* the mail. It is to arrive here day after tomorrow & scouts & pack animals will be left behind to bring on our portion. There will probably be no chance of sending letters E. after this & when next heard from it will be from Dufferin, whence I will probably telegraph on arrival.

I enclose a forget me not from the top of the Rocky Mts, of which I cannot furnish you the name.

Your loving brother

George.

{Many thanks for your Kamouraska Photo. You seem to be getting quite fat & strong again & the short locks are really quite becoming. GMD}

Skedan Indian Village, Queen Charlotte Islands, 1878

G.M. Dawson (third from left) and Party, Fort McLeod, B.C., 1879.

Looking up Waterton Lake, Alberta, from Lower End, 1881.

San Josef Bay, Northern Vancouver Island, B.C., 1885.

Indian Students, Alert Bay, B.C., 1885.

Boat Building Camp, Dease Lake, B.C., 1887.

Frances Lake, Yukon, 1887.

Looking up Pelly River, Yukon, 1887.

Kamloops, B.C., 1888.

Hydraulic Mine, Horsefly, B.C., 1894.

ENDURING ACHIEVEMENTS WITH THE GEOLOGICAL SURVEY OF CANADA

Dawson's geological work, from the time he became a member of the Geological Survey of Canada in 1875, was mainly carried on in western Canada. For many years he undertook field work in these largely unsettled and rugged regions. His geological explorations were of a very arduous character, and much physical energy and endurance were required to cope with the many hardships. For Dawson, it was a continual struggle against formidable distances and his own physical disadvantages. His Indian guides marvelled at his courage and strength, as indeed did all who worked for him. One who knew George Dawson well wrote: "He climbed, walked, and rode on horseback over more of Canada, than any other member of the Survey at that time – Yet to look at him, one would not think him capable of a day's hard physical labor. . . He could do as much field work as the biggest and strongest of his associates." It was his hand that first traced upon vacant maps the geological formations of the Yukon and much of British Columbia. "To all his expeditions, he brought a special kind of zest and leadership – this was a carrying on into more exacting circumstances of traits that had been noticed when he was just a boy playing outside the Principal's residence on the McGill campus. He was always cheerful, amusing, and popular, other boys flocked around him and invariably submitted to his unconscious leadership."

It was said that Dawson rode horseback exceedingly well, would take much care of his horses, and saw that others did also. He, if possible, would use the same horses throughout a summer's explorations, most unusual in such rough country. His favourite was a saddle horse named 'Major,' a sorrel with a heavy black tail and mane. "Major always came to camp at meal times for a lump of sugar."

In his explorations, Dawson visited such outlying regions of Canada as the Queen Charlotte Islands in 1878, north central British Columbia and the Peace River in 1879, and, in 1887, the Yukon. Our present knowledge of the geology of these districts is largely due to Dawson's observations at that time. White men and Indians still swear by his maps and geological data.

Without doubt, Dawson's major contributions to knowledge were made in the field of scholarly geology. His continuing concern for natural resources extraction was always substantiated by firm, theoretical geological study. Most remarkably, George's enormous physical energy and endurance were complimented by immense talents of observation and reasoning. As Morris Zaslow noted about George Dawson in his monumental study of the Geological Survey of Canada:

> *The acid test of any geologist is how well his work will stand up when it is re-examined later on the ground by a geologist equipped with new tools and informed by new scientific knowledge. Dawson meets this test better than any man of his generation. Those who have followed his paths and examined his notes and observations continually stand amazed at how rarely he reached an incorrect conclusion, how precise and acute were his powers of observation, literally overlooking nothing; also how much territory the little man covered in a single day, week or*

season. His powers of observation were matched by a keen intelligence; his generalizations showed the rare clarity, imagination, and originality of his mind. His ability to form sound and lasting general conclusions from a few reconnaissance observations and distant scannings was unique.[181]

Not surprisingly, even though refined and altered by subsequent sophisticated research, Dawson's conclusions about the complex rock formations of central British Columbia have in great measure been confirmed. His other great contribution which came with his first postulating the existence of a great Cordilleran glaciation which covered much of British Columbia, has similarly withstood the test of later study. His foundational glacial work indeed has formed the base upon which all later analysis has proceeded. George Dawson clearly did more than anyone else in the nineteenth century to advance the understanding of western Canadian, and especially British Columbia geology. The following excerpts vividly illustrate his thoughts on those two major themes of rock formations and glaciation.

"Geological Record of the Rocky Mountain Region," 72-73.

In my report for 1877 [Report of Progress, Geol. Surv. Can., 1877-'78] the existence in British Columbia of rocks shown by their fossils to be referable to the Triassic was made known, and these rocks, as developed in the Interior plateau region, were named the Nicola series or formation. This rests, at least in some places, unconformably upon the Carboniferous, and no rocks representing the Permean period have been identified. The Nicola formation is, however, chiefly composed of volcanic materials, the intercalated limestones or argillites in which fossils are occasionally found being few and far between. The greater part of its mass is undoubtedly Triassic, but the highest beds in a few places have yielded a small fauna that is referred by Professor Hyatt to the lower Jurassic. All the fossils are marine. [Fossils of the Triassic rocks of British Columbia, J.F. Whiteaves, Contributions to Canadian Palaeontology, vol. i, part 2. Annual Report, Geol. Surv. Can., vol. vii (N.S.), p. 49B et seq.]

Partial sections of the Nicola formation have been obtained in a number of places, but its study is attended with difficulty, owing to the very massive and uniform character of the most of its rocks, the region covered by it being best characterized as one of "greenstones." These rocks often closely resemble those of the Carboniferous, and in some places it is not easy to separate them, on the other hand, from the older Tertiary volcanic materials. Lithologically the rocks are chiefly altered diabases of green, gray, blackish, and purplish colors. In regard to state of aggregation, they comprise effusives (often amygdaloidal), agglomerates, and tuffs, the latter showing evidence of subaqueous deposition throughout the entire series. The tuffs are occasionally calcareous, and there are some thin and probably irregular beds of limestone, with infrequent layers of argillite. The most complete section so far obtained is one on the Thompson river, showing a total thicknesss of 13,590 feet; another near Nicola lake gives a probable minimum of 7,500 feet, and in both places more than nine-tenths of the whole is of volcanic origin.

The Nicola formation, with the characteristics above noted, is well developed in the central parts of the Interior plateau of British Columbia, and it probably extends far to the north in the same belt of country between the Coast and Gold ranges, but in the general absence of paleontological evidence, can not there as yet be separated, even locally, from the Paleozoic. [Annual Report, Geol. Surv. Can., vol. iii (N.S.), p33B.]

[181]Morris Zaslow, *Reading the Rocks: The Story of the Geological Survey of Canada 1842-1972* (Toronto: Macmillan Company of Canada, 1975), 112.

"On the Glaciation of the Northern Part of the Cordillera. . .," *American Geologist* 6 (1890), 154-55.

This part of the Cordillera of the West was, in the Glacial period, covered by a great confluent glacier-mass. Evidence of the existence of the southern part of this great ice-mass was, at an early stage in his [meaning Dawson's] investigation of the glacial phenomena of the region, obtained in the corresponding part of the interior plateau of British Columbia; and though doubts were at first entertained as to the mode by which the traces of a general, as distinct from the local, glaciation of the region might be explained, these were solved at a later date. [Quart. Journ. Geol. Soc., Vol. XXXVII, p. 283.] Still later, the writer was enabled, while engaged in an exploration in the Yukon district, to find evidence of the northwestward extension of the same confluent glacier and approximately to determine its limits in that direction. Having thus surrounded the area of this great glacier, it was proposed to name it the *Cordilleran Glacier* in order to distinguish it from the second and larger icecap by which the northeastern part of the continent was at the same period more or less completely covered. [Geol. Mag., August 1888.]

The Cordilleran glacier, as thus defined, had, when at its maximum development, a length of nearly 1200 miles. The main gathering-ground or *névé* of the *mer de glace* was contained approximately between the 55th and 59th parallels of north latitude, that part of the ice which flowed northwestward having a length beyond these limits of 350 miles, that which flowed in the opposite direction a length of about 600 miles. When at its greatest, a portion of its ice also passed off laterally by gaps transverse to the Coast ranges, and filling the wide valley between Vancouver island and the mainland, the ice there divided and flowed in opposite directions as the subsidiary, but yet large, glaciers of Queen-Charlotte sound and the Strait of Georgia. Ice from the main *mer de glace* does not appear to have crossed the Rocky Mountain range proper, on the other side, though considerable local glaciers were at the same time developed on the northeastward slopes of this range.

That portion of the Cordilleran glacier which moved south-eastward along the interior plateau of British Columbia, is now known, from numerous observed instances of striation crossing high points, to have covered the summits of isolated mountains of 7000 feet and over in hight; a circumstance which implies that the ice reached a general thickness of 2000 to 3000 feet above even the higher tracts of the plateau, while it must have attained a thickness of over 6000 feet above the main river-valleys and other principal depressions of the surface. [These statements depend in part on facts published in the Geological Magazine, August, 1889; in part on additonal evidence yet unpublished.]

The existence of this great Cordilleran glacier is naturally the first event of the period of glaciation of which evidence has been found in the region, as its ice-mass was competent to remove all signs of the more local growing glaciers which must have occurred during the early stages of the period of cold.

LIFE AND EXPLORATION ON THE WESTERN FRONTIER

It would not be possible to provide detailed accounts of Dawson's travels and explorations, but by reading some of George's letters, one can gain much insight into his work and adventures in these early days.

George Dawson to Anna Dawson, Victoria, B.C., 6 August 1875.

My Dear Anna,

Your letter from Metis[182] dated July 21 arrived today, & as a mail goes out from here tomorrow I sit down to report progress. I arrived here on Wednesday evening (day before yesterday) & have seen a little of the place, & hope to get off next week. The voyage from San Francisco was rather tedious, the distance being about 720 miles, & the steamer a very slow one. We had quite a number of passengers making Cabin & steerage about ninety – but little freight. On leaving San Francisco we passes through the 'Golden Gate' at once out on the Pacific ocean – which though moderately pacific yet favoured us with a long unpleasant swell. This continued increasing & for the first two days we had quite a little tossing with some rather cold wind & overcast dull weather. All hands were at once prostrated with sea sickness. At least I mean all the ladies, some gentlemen remaining <...> on deck. Personally I managed to Keep pretty well though with occasional unpleasant qualms. Our trip was quite a voyage of discovery, as neither the ship nor any of the officers had ever been to Vancouver before. The Pacific Mail Coy. to whose line the Salvador belongs have only just commenced running steamers here, having received the contract for carrying the Mails. Most of their vessels are on the Japan & Australian trades & {some} also run S. to Panama. The Captain a big burly, <ruff> rough old goose, thinking discretion &c. – bore quite away from the coast, & after loosing sight of the country near Frisco. we never rightly saw it again till the day we reached Victoria. Times on board were not particularly lively as the passenger list was not a remarkably select one. The food was very good, however, & what with reading a little, walking about the deck a great deal, & sleeping when these resources failed the time passed away. One morning I passed a great part of in the fruitless endeavour to fish up from the surface of the briny some peculiar Animals with which it was covered. The Animals were of the Jelly-fish nature, & looked like little 'portuguese men of war', having Some sort of floating body & a sort of hemispherical Crest rising up out of the water. In size they seemed to be from about 2 1/2 to 3 inches long & downwards, & as I never could see

[182]At Métis-sur-Mer, ten kilometres from Mont-Joli, Québec, on the St. Lawrence River where Dawson's parents built their summer home, "Birkenshaw," which still stands.

them very plainly they may have been *velellas* instead of *porpitas* for all I can tell. We sailed through them for about two & a half days. That is to say for that time you could seldom look overboard without seeing some of them. Often the water was quite dotted with them, & sometimes we would come across large patches of them nearly in contact. The Captain expressed his belief that there must be 'fourteen million' of them. We set to work to Catch them in a pail tied to a rope – I say we for several of the passengers enlisted in the attempt – At one end of the rope was an ordinary wooden pail, at the other end a small canvas bucket used for taking the temperature of the water. The experimenter used whichever of these he thought most hopeful, but we were going too fast, & I had now tow-net at hand, & even if I had had I do not think I could have used it. When we had quite satisfied ourselves we could not do it we left the apparatus which was then Seized on by a hopeful foreigner – German I think – who within a few minutes succeeded in loosing the whole affair overboard, & soon afterward the boy who washed up the deck cabins &c. was to be seen anxiously enquiring in all directions for his bucket.

The Crew of the Steamer, I should tell you were chiefly chinese, even to the 'Bosun', & so also were the waiters. They appeared remarkably handy & neat, & extremely bidable, not grumbling, & seeming to find a pleasure in their work. At the same time, they do not look strong, & I think there *was* a sort of Chinese flavour reminding in a distant way of opium, about the dishes.

I shall not be able to write to William this time, & fact I Suppose it will barely be worth while writing again to any English or Scotch address. If you write please let him know I am 'all right'. Also let me hear from him when you Can. I think it likely that where I am going on the Frazer River, I will be in contact with Surveying parties, & they have some sort of Mail communication weekly or fortnightly. However it takes a long time to get letters brought round even this far, as you may judge from the date of receipt of your own.

Your loving Brother

George

I am writing to Mother also by this Mail & addressing to Montreal, supposing that she will have got back there from Detroit before this arrives.

George Dawson to Anna Dawson, Blackwater River, B.C., 3 October 1875.

My Dear Anna

'Snow, snow beautiful Snow' at last; after a term of the most splendid cloudless weather, Frost nearly every night & oppressively hot in the Sun in the middle of the day. On Friday as we travelled along to this place the signs of bad weather began to appear. The barometer fell, light clouds began to gather & thicken. Geese flying from the far north passed overhead, & towards evening the wind went round from W. to N. & became squally. All yesterday it drizzled & occasionally a half melted snow flake Came down, & last night the snow Came in earnest & was this morning about 2 inches deep, Soft, & clammy. Today it is melting away as fast as it Can & the weather Shows signs of clearing up again, & I have no doubt that when this petulant outburst is fairly over, we Shall have the indian Summer for a week or two

I am now here waiting the arrival of a pack train from Quesnelle for Ft George,[183] Mr Smith[184] proposing that I should accompany it there & back the three pack animals I have had, having got So sore under the mismanagement of the indian packer, that they are for the

[183]Quesnel and Prince George.

[184]Marcus Smith (1815-1904) was the engineer in charge of the Canadian Pacific Railway Surveys in British Columbia.

present useless. The pack train may not get here for some days yet, but if the weather becomes settled I can occupy the time profitably in the neighbourhood.

I found quite a mail waiting my arrival here, the latest date being I think Aug. 23. Thanks for your letters & enclosures from William &c. I have not yet received a copy of my report,[185] but may yet get it before leaving here, & if not then on my return. The pack train goes to Ft George in 4 days & will not probably remain there more than a day or two. From here to Quesnelle is three days by pack train, though one can ride through light in a day, the distance being only about 45 miles. At Quesnelle of course one is on the regular travelled line of the Frazer & within less than a week of Victoria.

Nothing yet from Mr Selwyn, though I expect daily to hear of or see him. It is just possible that he may go out to the Coast by Lake François, of which {route} Mr Smith has written particulars to meet him at Ft George. In which Case I should not see him.

I think it is your turn for a letter, but for the life of me cannot remember to whom I wrote last. As there is nothing particular to write about however it cannot make much odds.

Since leaving the Cascade Mts.[186] there has been very little fine or remarkable scenery. The general surface of the country is a more or less regular plateau with hills & ranges of hills rising above it here & there, & river & stream Valleys cutting down into it. Some of the river vallies are very large & deep, but generally not rocky. There are occasional meadows, swamps, & hillsides with good grass, & in Some places open woods of large Douglas pines[187] with grass growing beneath them. As a rule however the country is pretty thickly wooded with Scrub pine,[188] Sometimes of fair growth & standing, often prostrate, often burnt, often burnt & prostrate with young scrub pines coming up in dense thickets &c. &c. Through & over & among all these varieties of country, the little paths they Call trails here, wind & twist seeking for lines of least resistance, or following the <path> track by which the first indian originally scrambled across the country. I think I once used the Simile before, but really the country much resembles a gigantic game of spillicans,[189] & I believe if one was to go to Ft George & begin shaking the pile, sticks would be Seen moving down at New Westminster on the Frazer!

There is Something very pleasant however in these autumn woods, now that the aspens are turning yellow among the spruces & pines. There are no maples however or other trees which take really beautiful tints.

I am progressing rapidly in my knowledge of Chinook,[190] & can now understand pretty well & Speak a little. It is very easy to learn of course there being no grammar at all & a few words going a long way, especially when the range of subjects usually forming topics of conversation is limited.

I am writing a few words to Father, & May have time for another letter before leaving here. If not it will not now be long before I get back to more regular postal communications.

With love to yourself & all

[185]Dawson was referring to the report of his work on the boundary commission published as his, *Geology and Resources.*

[186]Coast Mountains.

[187]Douglas-fir, *Pseudotsuga menziesii* var. *glauca* (Beissn.) Franco.

[188]Lodgepole pine, *Pinus contorta* var. *latifolia* Engelm.

[189]A game played with a heap of small rods of wood, bone, or the like, the object being to pull off each by means of a hook without disturbing the rest.

[190]A trade language widely used by Northwest Coast Indians.

George Dawson to Anna Dawson, Victoria, B.C., 14 November 1875.

My Dear Anna,

{I have not written to W. for Some time. Please tell him I will answer soon. If this family gets much more scattered one will have time for nothing else but writing.}

Many thanks for your last long letter received a day or two ago, & first to answer your question as to the proper times for mailing letters – The mail steamers Sail from San Fran. on the 10th 20 & 30 of Each month, & the newspapers are kept back for them. Letters generally come overland arriving once to twice a week. The time of transit overland or by steamer is about the Same.

I had stored up quite a lot of information about our journey down the Frazer & waggon road to this place, but not having had opportunity to write it on first arrival here, have almost forgotten it all. From Ft George to Quesnelle, a distance of about 80 miles, we came down the river in a Canoe & boat; the same which Mr Selwyn had with him on his long trip. The stream is very <rapid> {swift}, & full of little riffles or small rapids which generally require to be run with Some Care. There are also two 'Cañons' so called, with larger rapids in them, but these did not turn out so formidable as we had thought, the water being low. The voyage occupied 2 days & a half, & it blew a Gale all one day dead ahead. It must have been a remarkably heavy gale, for during its height we Could hear & see trees Crashing down in the woods in all directions. on one occasion we tried to cross the stream but drifted down into the wrong Channel of a rapid, where the current was very boisterous & the waves were all turned backward & their tops blowing off with the wind. Just as we got the Canoe out of this into dead water at the foot of the cliff, a great pine almost directly above our heads blew over, but was fortunately prevented from coming down on us by Some smaller trees which did not give way. We had to wait Several days at Quesnelle, which is a dismal place, for the Steamer; when our afternooons steaming brought us to Soda Creek. Here we got the stage, & after four long days staging – from before dawn till long after dark, – arrived at Yale. The only noteworthy Circumstance in this part of the journey was an upset which we were unlucky enough to meet with. This took place by the driver getting off the road & onto the sloping side of a bank. Mr S & I were Sitting on the box, & as it was extremely dark I cant' tell you exactly what attitudes we assumed in falling, though if I may be allowed to judge by the appearance of my hat, I Should be of the opinion that I fell at least partly on my head. The other passengers were tightly buttoned up inside, & as soon as they could make an orifice they emerged hurridly, one by one. Last Came Mr Walkem[191] the premier of this Great Province, much crushed & speechless. He declares that all the rest fell on him, & then scrambled on top of him to get out. Fortunately the driver kept the reins, & the horses stood Still till they were detached. Such are the pleasures of travel in this country. No bad accident has ever happened to the stage on the Frazer R. road which almost Seems miraculous when one sees how the drivers come Skimming down the long hills cut in the sides of the cliffs & mountains, & sometimes actually overhanging the river below.

Oct. 15.[192]

Here I stopped last night, & now take a few moments to add a line or two.

First as to the question on which you ask advice in your last, *viz*, the house. Now I do not know how at this distance you can expect me to give a rational opinion on the subject, So much seems to depend on circumstances, & especially on your own opinion & wishes. I am afraid to give any opinion in Case it <might> {may} influence you where your own judgement Must be so much better. At the same time it does Seem to me that so large a house, & one especially with such large rooms *might* prove rather a burden than otherwise, & would be

[191]George Anthony Walkem (1834-1908) was premier of British Columbia from 1874 to 1876 and again from 1878 until 1882.

[192]The date should read 15 November.

uncomfortably large for two people to inhabit.

I dont' Know how it is but when I sit down to write my thoughts never flow freely enough; & I cannot think of any quantity of nice things to Say as you do. I dont' think I am reserved as you say, but only unfortunate in not being able to catch the right idea at the right time.

Your loving Brother

George

Please tell Father that I have answered Capt. Featherstonehaugh's question as to the height of W. Butte.[193] Also that I hope to write soon. Also congratulate O'Hara for me on his success, & tell Rankine that If Mr Selwyn gives him a Cartridge, I hope he will keep it, till I write & tell him what I want done.

The proof photo received some time since was nearly black from action of light. It seems good however what can be seen of it, & I long to have a decent Copy.

George Dawson to Anna Dawson, Victoria, B.C., 20 December 1875.

My Dear Anna,

I have been writing So Many letters last night & today that I feel rather tired of it, but must not let the steamer Sail without Sending something home. Mails go twice a week by "the Sound"[194] & thence overland to Sacremento, but the steamer Saves several days in time.

Your letters acknowledging receipt of my first notes from Victoria since my return have only arrived a few days ago – So long does it take to question & answer from this desolate isle of the sea. I am sure I *meant* to say everything nice with regard to your engagement, So I am glad that you Kindly refrained from your first idea of scolding me. Joking apart you know I do not "gush" on paper, but I am really delighted, & if I Could only think of what to say & how to say it should write to Harrington himself & tell him so, or rather congratulate him on his good fortune.

I have sent off by express to the Geol. Survey a number of boxes today, & in one of them have enclosed, addressed to Father, a small thin parcel. It Contains two Photos. one of Victoria the great western Metropolis, the second of Yale on the Frazer R, where that stream leaves the Cascade Mts. Also a sketch of my own, the only one I have made this summer which I consider at all passable, please take the photos or the picture whichever you like best. The Sketch is of Tatlayoco Lake where it runs up among the Cascade Mts. The proposed railway route passes along the bank to the left to the bottom of the lake, then crosses the head of the Homathco R which runs out of it, & follows its valley (behind the range of hills fringing the lake on the right) to the Sea. That is if the ry. is ever built it will go <here> thus, & I hope those fools at Ottawa are not going to throw away their last chance of consolidating the Dominion because there happens to be a commercial depression in the year of grace 1875.

The boxes will probably not arrive till about a week after this letter.

I fear this letter will be too late to wish you a merry Christmas, but if it does as it should, it ought to at least to arrive Christmas week, & allow me to wish you & all at home a Merry New Year, which I hereby do with much earnestness.

I can hardly believe that it is so near Christmas as it is, the time seems to have slipped away So fast, & left nothing to show for it, & then to there is no proper winter to mark the time, only a prolonged & dismal Autumn, with the vine still clinging to the Mouldering wall & the rain & the wind are never weary. The grass is green & fresh looking, as it should be if water will make it So, & hardy vegetables

[193]Captain Albany Featherstonhaugh, assistant astronomer on the boundary commission, wrote to J.W. Dawson to clarify an apparent discrepancy in the altitude measurements made by George and the Americans.

[194]Mail sent across Puget Sound.

still Stand out in the fields; the thermometer only occasionally reaches the freezing point at night. The arbutus, an evergreen tree, is green in the woods, where the ground is also covered with mahonia bushes & Sal-lal plants – A species of *Gaultheria*.[195] Now too the Moss which in summer is dried to a crisp & choked with dust, & looks altogether out of place; is washed Clean again & covers rocks branches & shingle roofs like so much wet sponge. The woods are uncommonly full of moss here, & not only underfoot, for it creeps up the tree trunks & settles in tufts & cushions even on the exposed branches. The Maple trees especially are generally Shaggy with moss of various colours, & now & then you may see a *polypodium*[196] perched up aloft in it. Now to stray seeds which in summer have become entangled in the Moss begin to sprout & you may see forests of little pines in some places growing up even on the trunks of the older ones. Where stalks of weeds have fallen, the Seeds are striking out even in the Maternal pod, under the influence of the damp.

I have an invitation to dine on Christmass with Dr Helmckin[197] (I hope that may chance to be the right spelling) It is very kind of the good man, & I hope I may appreciate it properly. He is a most excentric genius, & goes about with a great cloak of which the apperture Slews round in all directions but is never exactly in front. He has one grown up daughter, who – Shall I say it? – well she appears to squint, &, for your peace of mind be it added, is supposed to be engaged to a certain Mr N – after Such a description names Must be under the seal of – friendship.[198]

With love to all

Your affectionate

George Dawson to Anna Dawson, Victoria, B.C., 6 January 1876.

My Dear Anna

I must thank you very much for your very acceptable present of a portable photo. album. Your photo. is very good, I think, fully as good as any you have had taken before. I have not shown it to the Creases,[199] nor did I dine there on Christmas day as you suppose in your last letter. The C's are unfortunately in trouble just now. The second daughter while down at Esquimault paying a visit took Scarlet fever, & a day or two afterwards all those at home took measles, some of them very badly. I believe they have been unable to get servants & have had quite a time of it generally.

Now for All the other presents I have to thank the doners through you. Rankine for his Capital diary, & Eva for the peculiar cork pen-handle. There is nothing new here to report & things seem to have settled down pretty much into winter train. I am not working hard as you fear but taking life remarkably easy, & rusting slowly. Two or three years in this dead & alive place would cause anyone to loose all idea of the value of time, & quite spoil one for life anywhere else. Messrs Jennings,[200] Cambie,[201] & Harris,[202] three leaders of parties on

[195]The "arbutus" was arbutus, *Arbutus menziesii* Pursh; "mahonia bushes" were tall Oregon grapes, *Berbenis aquifolium* Pursh; and "*Gaultheria*" salal, *Gaultheria shallon* Pursh.

[196]The "Maple trees" were probably Douglas maples, *Acer glabrum* Torr.; and "*polypodium*" licorice fern, *Polypodium vulgare* L.

[197]Dr. John Sebastian Helmcken (1824-1920) was one of Victoria's best-known residents who served as medical officer and later surgeon to the Hudson's Bay Company from his arrival in March 1850. He was also heavily involved in politics and was one of three delegates sent to Ottawa to negotiate British Columbia's entry into Confederation in 1870.

[198]Catherine Amelia Helmcken, who was then twenty, did not marry "Mr N" but rather George Archibald McTavish on 4 December 1877.

[199]Sir Henry Pering Pellew Crease (1823-1905) and his wife Sarah (Lindley) Crease (1826-1922) were prominent Victorians with whom Dawson socialized frequently in the winter of 1875-76. Sir Henry had earlier served as attorney general of British Columbia before being made a puisne judge in 1870. He played an important role in British Columbia's entry into Confederation, drafting the Terms of Union in 1870.

[200]W.T. Jennings was in charge of one of the divisions of the Canadian Pacific Railway Surveys.

[201]Henry John Cambie (1836-1928) was in charge of one of the Canadian Pacific Railway Surveys parties in British Columbia and later superintended the Canadian Pacific Railway's construction in the province from 1880 to 1883.

[202]Dennis Reginald Harris, (1851-1932), a civil engineer, was employed by the railway surveys.

the C.P.R.S. left for Ottawa a day or two ago. Mr Marcus Smith, & Gamsby[203] leave next boat, & then there will only be a few of the Subs. here. As for myself I dont think anything will induce me to spend another winter here, the most out-of-the-way Hudson's Bay post on the Continent would be preferable. The weather continues moist moister & moistest, though I must say we have had two fine days lately, & they were really very fine, & quite warm. Rightly concluding on both occasions that the weather looked Settled, I sallied forth to do a little Geology in the neighbourhood, took some lunch with me & worked all day along the shore, finding it almost hot in some places. One day there was a tremendous Surf rolling in & breaking against the rocks. I do not remember before to have seen such great waves Coming ashore, the spray flew high into the air, & went scudding across the grass. There are some beautiful spots along the sea shore here, at least they must be beautiful when the trees have leaves on them & the ground is not in the State of a quagmire. Sloping lawns naturally planted with scattered oaks[204] & bushes, & splendid views of the Snow Covered olympian Mts. Mt Baker, the highest peak in Sight from here I have only seen once, & that in August last. On one occasion the smoke from burning woods cleared away enough to show it, but never since my return here have the Clouds forsaken it.

Your loving Brother

George Dawson to Anna Dawson, Victoria, B.C., 21 February 1876.

My dear Anna,

When I last wrote to Father our condition here as to Mails was deplorable. It is even more so now there having been no relief. The last Canadian letters are now a Month old by their dates, & we cannot expect any newspapers for at least a week as they only come by the direct steamers once a fortnight, & the last steamer brought none. As for the letters I do not Know whether they are snowed up on the Union Pacific Ry, or somewhere north of Calafornia.

Here the weather is much the same as ever. Hardly a day without rain, but occasionally quite warm & Summer like, with birds singing & the willow Catkins beginning to come out. Then probably in a short time it will be blowing a gale with rain & perhaps sleet, & quite uncomfortably Cold.

The town today is in a great state of excitement over the election, the new Premier Elliot [205] being declared head of the Pole after a pretty hard Contest. All the Elliotites are even now rushing about & congratulating each other, while the beaten party are doing the reverse whatever that may be, & both sides are indulging to the benefit of the Canadian revenue, in various exciseable articles.

I have Almost finished my Survey report,[206] though unable to post it today. Tomorrow I hope to see it through, which will be a bless-

[203]C.H. Gamsby was an engineer in charge with the Canadian Pacific Railway Surveys.

[204]Garry oak, *Quercus garryana* Dougl.

[205]Andrew Charles Elliott (1829?-1889) remained premier until 1878 when he lost his seat in the legislature.

[206]The report of Dawson's 1875 field work was delayed and published later as, "Report on Explorations in British Columbia," Geological Survey of Canada, *Report of Progress for the Year 1875-76* (Montreal: Dawson Brothers, 1877).

ing, as I will then be able to get about again a little & do out door work in connection with the geology of this place. I hope Soon too to get the last of the Exhibition[207] things away which will be another Cause of rejoicing. About the first of April from all I hear it may be possible to get out in the field in the Island or along the Mainland Coast without fear of being deluged or Snowed upon, & when this occurs it will be a third cause of delight, & infinitely preferable to living here in Victoria. If one can begin in April & go on till October it Should allow a pretty good Summer's work to be done, & that fairly over I shall be ready to return to Montreal for a Space.

This is a very unsatisfactory letter I feel, especially as it was intended as an answer to the long one you wrote me giving an account of your Visit to the Harrington Family. Still there seems to be no proper head of steam on. So having Netted wearily through the mental water for some time without so much as Capturing a prawn, I think I must even give up for the present & go & post this, taking a walk at the same time among the free & independent electors of Eatansmill.[208] Enclosed is a leaf of a Japanese [Punch] for Eva instead of a letter. If she can make out what it all means I shall be glad to know

Your loving Brother

George Dawson to Anna Dawson, Victoria, B.C., 30 March 1876.

My dear Anna,

I have had your picture of Montreal from the Mountain, now for some time, & have been daily intending to thank you for it by letter. If the second copy is as you Say <as go> better than the first it must indeed be good, as the one I have gives the idea remarkably well – exactly as I remember it. I also have your long letter – I forget the date – but about March first – By the same mail I had another long letter from Ella – Mrs Kemp[209] – enclosing a Sort of Christmas Card, but beautifully painted by herself with rosebuds forget-me-nots, & dear Knows what besides. I must answer her soon.

No one has ever yet told me what like the photographs from Ottawa – Commission views – were, whether satisfactory or not, & how many. Dr Burgess writes that he has just received his. He also informs me that he has an "heiress"[210] I daresay you remember he was married last winter. Quote the – what you are pleased to Call general – "honey moonish" aspect of the news in your letter Things seem to be taking a similar turn here, if one may judge by the general tenor of the conversation of the people. This may partly arise however from their having nothing else to talk about. I do not *particularly* object to your sermons as you seem to think I do, except in so far as they occupy space which might be devoted to other objects. Believing as you do, you cannot help feeling as you say, but I am sure you would be the last person in the world to ask anyone to *try* to make themselves believe in that which does not appear reasonable to them; for this would be mental dishonesty just as bad on one side, as if you, believing as *you* do were to try to force yourself to think something else. – But I tread on dangerous ground, for I feel what I write will be weighed, & if wrong impressions are conveyed they Cannot at once be rectified. I hope all you *think may be* true on *another* subject, is not, for I feel

[207]The United States Centennial Exhibition in Philadelphia was a major international event. Dawson had resonsibility of collecting both for the Geological Survey of Canada whose collection was meant to be a representative sample of the geology and mineral resources of Canada, and for the broader Canadian collection.

[208]Eatansmill was the scene of the corrupt election portrayed by Charles Dickens in *The Posthumous Papers of the Pickwick Club* (1836-37).

[209]Mrs. William Kemp, of Leith, Scotland.

[210]In 1875, Dr. T.J.W. Burgess, lecturer at McGill and former colleague of Dawson's on the boundary commission, married Jessie MacPherson of Whitby, Ontario.

unable to take exactly the same view, & fear I never shall. – But here again I leave the solid ground & must flounder onto terra firma by assuring you of <my> {the} constant remembrance of

Your affectionate Brother

George

{Evas letter is received. Please thank her for it & give her my love}

George Dawson to Anna Harrington, Victoria, B.C., 13 May 1877.

My dear Anna,

As already announced in a short note to Mother, I arrived here on Thursday Evening last, my previous movements should also have been pretty well defined if the post cards dropped from time to time have all reached Safely.

I suppose you will expect to receive some Short account of my journey, which occupied sixteen days, & so will try to give you a few notes. – The time occupied would have been just 14 days but for two days of forced rest, one at Roseburg on a Sunday, the second at Seattle on "the Sound" waiting for the steamer, which only makes two trips a week each way.

At Port Huron I got safely through the hands of the Customs officers, without any Particular trouble, & proceeded on to Chicago in the regular course. Thence changing cars, & {after} waiting about an hour onward to Omaha, arriving there in a great storm of wind & rain – by no means pleasant as two changes have to be made, the first into the Bridge transfer Cars, & then after Crossing again into the Pullman on the U.P. Ry.[211] In the western states, & at Omaha every thing was green, somewhat in advance of Montreal, willows being in flower &c., but on going a few miles further west, vegetation being at about the same stage, we got into a snowstorm, which continued with little intermssion till we reached the Summit at Sherman. Several miles befor Sherman we got into quite deep dry drifted snow-banks & had some difficulty in getting through with three engines & a snow Plough. On leaving Omaha people know that they are bound to spend a few days together as on a short sea voyage & in Consequence at once become more conversational & friendly. There were two Pullmans on the train & some very pleasant people most of whom I forget already. I had a seat with a Mr Wise, connected with the Ry. office at Ogden Utah who proved intelligent. Near at hand was an Americanized Dutchman appointed as Postmaster to some remote district, & travelling with his daughter a not too ugly girl but too modest to speak. There was a Mrs Lawrence from Washington going out to join her husband, a doctor in Arizona. Also another lady who soon became a chum of her but got off to go to Gold Hill in Nevada, where her husband is something in mines – Also a number of other people, including several young ladies all of whom with the exception of one were sufficiently plain looking & common place in every respect. The one was returning to Oakland in California from a visit East, with her mother & other members of the family – she was extremely good looking to say the least of it & clever also – but as I can just imagine you running your eye along the next line to see all about it I shall pull up here without telling you her name or anything else. We had a small Cabinet organ on one of the Pullmans, & the musical portion of the community continued to enliven the journey with songs till we lost the music box in changing cars at Ogden. About the only tune everybody seemed to know was *Hold the Fort*, so this & some others of M. & S.[212] series were favourites. Disgusted with the loss of our instrument we solaced ourselves by sitting out on the platforms & steps waiting for sunrises, sunsets, meridian passages & other astronomical phenomena – watching them, & then going back indoors to shake bushels of dust out of our clothes, & wash pebbles out of our eyes. One sunset over the Great Salt Lake was very satisfactorally observed.

[211]Union Pacific Railway.

[212]Dwight L. Moody and Ira Sankey were American evangelists who produced many widely-used hymns.

We also observed antelope which appeared to reciprocate in most cases, & from time to time several people would dash themselves violently against a window to see a Jack rabbit which never could be seen. In the Salt Lake Valley we found apples, peaches & &c. in full bloom & everything like early Summer, but soon ran into the wastes of the Humboldt Valley, & thence Crept up to the summit of the Sierra Nevada where the appearance was that of very early spring with plenty snow lying about. From the summit {which} you attain in the very early morning, you run down to San Francisco in one day, arriving there in the evening. As we slid along, – the only use of the engine now being to keep the train from running too fast – we Passed down through the foot hills now green & beautiful with flowers to the wide Sacramento Valley, & found ourselves in Summer, haying going on strawberries ripe & all the trees in full foliage. At Roseville Junction near Sacramento I reluctantly left the comforts of the Pullman, & in five minutes found myself alone & forlorn kicking my heels in as dull a little country waiting room as you ever saw. I had several hours to do it in & so did not hurry, but tried reading & walking & having some dinner, & at last four o'clock Came & with it the train for Maryville, surcharge with hot local passengers intersprinkled with babies. At Marysville I had to get a new ticket, telegraph for a place on the stage, get luggage rechecked & supper within 20 or 25 minutes – which being fulfilled we jogged on to Reading, stopping at all Sorts of little stations by the way, but finally pulling up at the terminus at about 1 a.m. After going to bed we got up at half past four to breakfast for the stage starting at five. The first view of the stage was not very reassuring as the inside was half filled with mail matter, & a miscellany strapped all over the back top. For Passengers we had a stout Great grandfather & his wife going up to Oregon to see their children, an Americo-German Jew buying skins & furs, an attorney to something & a non-descript man also with a wife. We gradually settled as the stage moved on, those at first occupying the honourable but incommodious position of keystones, finding themselves on the seat in the course of half an hour. The great grandmother would have kept plenty of room in her corner had she been strong enough, but fortunately years had done their work & she was'nt. Various kalidescopic changes occured as we went along which even to remember would be tedious. We lost some passengers & gained others, till finally the losses preponderating I found myself alone with the driver on arriving at Roseburg. From Reading we drove on for three days & two nights, the longest stop we made being at Yreka for two hours. The roads were execrable having just hardened after the winter rains, & being composed of ruts & hard intervening ridges, in most places so narrow that it is with the greatest difficulty a place can be selected for two coaches to pass. Though not passing along sheer cliffs as on the Frazer River, this road cut out for long distances like a shelf on steep grassy hill sides & with almost impossibly steep hills to ascend & descend, is really far more dangerous. At one place near the north line of California you ascend the Siskiyou Mountain – a sort of pass – over 2000 feet high by means of a system of most involved doubling too & fro. On the top, you at once begin to descend again by a road very like the ascent I suppose though I dont' know much about it as it was pitch dark. I only heard the brake shrieking against the wheels as we went bumping along, & saw the horses (six of them) apparently dancing on the edge of an abyss as we flew round the curves. However we reached the bottom at last & I dont' particularly want ever to see that hill again. One begins to get sleepy too about the third day. You are admiring the scenery – paying the greatest *possible* attention to it – when all at once you relapse into a state of temporary insanity with the most absurd dreams rushing through your head till <all at once> {suddenly} you wake just on the point of jolting forward among the horses. When at last we stopped I found myself all covered with contusions & tender spots, hands brightly polished with holding on to the iron rails, & head nearly sawn off by the edges of my collar, but still in a capital condition for a good night's sleep. Next day we went on again in the stage till about 3 P.m. when we reached Roseburg. Thence to Portland Ogn. by rail, thence on the Willamette & Columbia by steamer to Kalama, thence by rail to Tacoma, thence by steamer to Seattle & finally thence by steamer here. On reaching, at last, the Pacific at the lower end of the Sound they presented us among other delicacies with clams. I proposed to eat some of these in honour of the occasion & began on one, but finally concluded I would try Something else as we had only half an hour for supper. I cannot enter into detail however with the various gastronomic struggles experienced during the journey, or the varying consequent forms of indegestion. It would be too sad a subject to close with. I find too I have not said anything about the scenery which was the chief object of the letter at the outset, also that my record of the weather has been prematurely brought to a close about page 2. both of which circumstances you will no doubt deeply regret. Some of the country is really very pretty, & it is all well worth seeing once. You pass in Northern California within a few miles of the base of Mt Shasta a wonderful snow clad volcanic mountain over 14,000 feet in height, & almost isolated. Then the Valley of the Rogue River, & those of the Shasta, Klamath, & Umpqua are remarkably beautiful. The country is generally mountainous but with little rock showing, bare grassy slopes rising steeply from the valleys to heights of 1000 or at least several hundred feet, scattered clumps & groves of fine well grown & rounded oaks, maples, & tall firs. The whole giving the effect of a perfectly kept park, with wide fertile fields in the flat bottom-lands of the valleys, all at this season beautifully green. The climate too is probably as fine as any in the world, with scarcely anything that we would call winter, less rain than on the actual coast, & an Italian Summer admitting of the easy cultivation of grapes on the large scale. It being Sunday I may be allowed to characterize it in the words of the psalmist as a region "Where every prospect pleases, but only man is Vile" for really I never heard so much concentrated bad language before as during the last ten days.

The Steamer from San Francisco is expected in hourly & by it I hope to hear from home & to know that your convalescence has been complete. I feel already as if I had been journeying forty years in the wilderness since leaving Montreal, but hope to enter into the Promised Land via New Westminster about next Friday. I fear I have written a great deal of nonsense & so shant read it over to see. I address to McGill as I have forgotten your number please send it to me on a slip.

Your affectionate Brother

Now, in May 1878, Dawson again arrived in Victoria after a long trip from Montreal. This season he sailed to the Queen Charlotte Islands to investigate geology, and the customs and habits of the Haida Indians.

George Dawson to Anna Harrington, Victoria, B.C., 19 May 1878.

My dear Anna

I still date, worse luck, from this place, but hope to get away this week, though the precise day is yet uncertain. When we got here, after a remarkably quick & pleasant passage from San Francisco, Capt Douglas & his Craft were not to be heard of, but as I had given him till the 15th I was obliged to wait till that date, making enquiries in the meantime as to other available Chances. On the 16th I made an arrangement for the schooner *Wanderer* of about 20 tons & as soon as she is ready for sea will be off. She belongs to a man Called Sabotson[213] who is getting new sails for her <...> & as Soon as these are ready all will be complete. Since engaging the Wanderer I have heard that Douglas' schooner which Mr Richardson[214] formerly had was blown ashore in a squall somewhere near Comox & so much damaged that she Cannot be put in repair for some time. The *Wanderer* is I think a good Craft, with plenty beam, & built originally for a Pilot boat. If the Captain & Crew, the latter Yet to be selected, turn out well all will be right.

There are as you may suppose innumerable little things to look

[213]John Sabiston (1853?-1898) who captained the *Wanderer*, Dawson's schooner during the 1878 field work in the Queen Charlotte Islands, later died of brain disease in New Westminster.

[214]James Richardson (1810-1883) was a geologist with the Geological Survey of Canada for some thirty-six years, whose early 1870s explorations formed the basis of his pioneer work on British Columbia coal fields.

after here & it would seem that we Cannot even get a clear half day for any work about this place. Tomorrow I shall make a vigorous effort to clear off the remainder of my list, so that any days remaining may be free from little engagements Rankine Comments on the fewness of good looking girls in this part of the world, especially in contrast with their abundance in San Francisco. The best looking & therefore most highly esteemed he has yet encountered is a Miss McDonald,[215] daughter of the senator of the same name[216] who travelled over the ry. with us & asked us to his home the other night to dinner.

The weather is fine & settled looking now, though there have been a few windy Cold days since we landed. I long to be off taking advantage of the fine time. When all is ready, following advice, I shall strike straight away for the Queen Charlottes & after spending as much of the summer there as the Country seems to warrent & weather permits, shall begin to work back toward Vancouver. I enclose a little memo. for Mamma.

Your loving brother

George Dawson to Anna Harrington, Victoria, B.C., 26 May 1879.

My dear Anna,

We arrive here yesterday morning at half past four <AM> but being the Queen's birthday found everything shut up & the town in fact nearly deserted, everyone going off on excursions of one sort or other. Today being Sunday continues the enforced idleness but on Monday I hope to get much done toward moving on. It is, however, scarcely likely that we can leave here for some days yet, & indeed it will not advance us to do so as our transport, in the form of pack mules &c. cannot yet for some time be ready to set out from Quesnel. As already advised it is probable that I may go up the Skeena with some of the railway engineers meeting the trains at Fort St. James, Stuart Lake. When all plans are fully made I will write again, giving details. As far as I can see at present, however, it will be little use sending letters here after <...> the middle of next month. In fact it is very doubtful whether after leaving here we shall hear from home till some time next October. There may be chances of sending letters out till about the first of July, but after that date probably none. I hope no one at home will be anxious if letters are not received.

On the way across I have gone through most of the Qugoldsby legends & shall mail the book back to you in a few days.

We came up from San Francisco on a steamer called the "City Of Chester" comfortable enough, but a terrible roller, which circumstance rendered us all rather pensive for a day or so. However we got well over this & enjoyed the latter part of the trip. Besides Myself & McConnell[217] there were several CP Ry engineers &c. *viz*. Cambie, Keefer,[218] Horetzky,[219] McLeod &c. with the Rev. Mr Gordon[220] from

[215]Probably Flora Alexandrina (1867-1924) the eldest of three daughters.

[216]Senator William John Macdonald (1829-1916) was a British Columbia businessman and politician who was one of three British Columbia senators from 1871 until 1915.

[217]Richard George McConnell (1857-1942) had just graduated from McGill and was to spend some thirty years with the Geological Survey of Canada. McConnell later accompanied Dawson on several other expeditions and rose through the survey ranks until he became deputy minister of mines in 1914, a position he held until retirement in 1921.

[218]George Alexander Keefer (1836-1912) was a surveyor and engineer with the Canadian Pacific Railway Surveys who later remained with the Canadian Pacific Railway until 1886, supervising construction of the difficult Fraser Canyon section of the line.

[219]Charles George Horetzky (1839-1900) was a photographer with the railway surveys who took many invaluable photographs of British Columbia topography but was extremely difficult to work with.

[220]Daniel Miner Gordon (1845-1925) was a Presbyterian minister who later served as professor of systematic theology at Halifax Presbyterian College then principal of Queen's University from 1902 to 1917. Gordon's account of his 1879 journey with Dawson was subsequently published as *Mountain and Prairie*. . . (Montreal: Dawson, 1880).

Ottawa who intends with Mr Fleming's[221] permission accompanying one of the engineering parties across to Red River. He has doffed his clerical rigging for the occasion. This scribble written with bad pen ink & paper at a desk in the hotel is merely to let you know of my safe arrival by the earliest mail. When feeling more inclined for the business & with better opportunities I shall write more fully

Meanwhile believe me, with love to all

Your affectionate brother

George Dawson to Anna Harrington, Stuart Lake, B.C., 6 July 1879.

My dear Anna

Your long letter of May 19th was handed to me this morning, we having arrived here last night between 10 &11 o'clock. I did not expect any letters, as the date at which they might leave Victoria was scarcely later than that at which we ourselves started for the Skeena. We have now formed the long anticipated junction with the pack-trains from Kamloops, & today being sunday we are making a day of rest, hoping to be all ready to make a new departure on Tuesday morning. Monday will be occupied in arranging about pack-trains & getting the loads &c. adjusted.

Since writing last from the Forks of the Skeena, we have crossed the "portage" thence to the north end of Babine Lake – some 50 miles in length; ascended the lake to its head – 100 miles – Crossed another portage to the head of Stuart Lake – 8 miles –, & come down to the outlet of the lake, a distance of 36 miles. The so-called portage to the lower or north end of Babine Lake is a trail through a rough mountainous country, with high snowy peaks projecting on all sides. We started from the Forks on Monday morning about 9 A.m., with a train of 22 Indians, – two of whom were women – carrying about 100 lbs each, the load being tied into as compact a bundle as possible & secured by a strap across the forehead & chest. We had fine weather on the whole, though rather warm, & climbing up & down the steep rocky hills & through the woods was fatiguing enough to those of us who walked light, & as you may imagine brought the perspiration out on the packers. We were four & a half days making the distance, as it was impossible to get the Indians along at a greater pace than about a mile an hour, including rests, which were frequent, & almost always entailed the necessity of lighting a smudge to keep the mosquitoes & black-flies at bay. You would have laughed could you have seen us getting out of Camp in the mornings. The waking call at 4.30 or 5 o'c, stowing away blankets &c, washing & dressing; breakfast, & then the feeding of the Indian packers who generally sat in a circle & had their food doled out by one of themselves who was appointed cook. The loads had then to be baled up & the caravan put in motion, the Indians generally packing the little tents they brought to sleep under, on their dogs to lighten their own loads. We generally remained an hour or two behind to let them get a fair start, & then followed, resting an hour or two again in the middle of the day. The greatest altitude we reached was between 4000 & 5000 feet, but that was scarcely above the timber line & the higher snow clad ranges & peaks looked higher when thus seen from some little elevation, than when looked at from the valleys. Some of the mountains probably surpassed 8000 feet, & about the shady sides of their summits a few small glaciers lurked. On Friday, about noon we camped on the shore of the north end of Babine Lake, in a downpour of rain, the camping ground being

[221]Sir Sandford Fleming (1827-1915) was Canada's foremost railway surveyor and construction engineer of the nineteenth century, and also a distinguished inventor and scientist. Fleming had been appointed engineer of the proposed new railway to the Pacific coast in 1871, and was in charge of the surveys across the west.

the edge of a swamp. The Indians of the Babine village seeing that we required canoes to go down the lake but up the price of canoes & men to a fabulous figure, & only after having engaged the chief's canoe at an exhorbitant rate were other prices lowered. All saturday it rained & blew & it was not much better on Sunday, so that we did not get away till Monday morning. Babine Lake is about a hundred miles long & probably eight miles wide in some places. The first day we ran about 25 miles, camping near another Indian village & small H.B. post. Here Mr Cambie & Macleod wished to examine a possible pass eastward & to remain a day, so Mr Gordon & myself in the second canoe went on <Camp> reaching the head of the lake on Thursday, about noon. The weather was on the whole remarkably fine & the scenery beautiful, while we caught with spoon-baits trolling behind the canoe more trout than we could eat. On Thursday, however, when we reached the head of the lake it was raining by pail fulls, & I had some trouble with the Indians in getting them to shoulder the stuff for the portage. We could get only one man to help the crew & were obliged to leave some provisions to be brought on by Mr Cambie. As it was the loads the men carried were very heavy & they grumbled considerably. We camped on the portage trail, wet & surrounded by a storm of mosquitoes, & about 4.30 the next morning reached the head of Stuart Lake. I sent the men back at once to meet Mr Cambie & help to bring his stuff across. They met the other party half way over the trail & in the afternoon we were all together again. It had been expected that a boat would be sent up from Ft St James to the head of the lake on the arrival of the pack trains at the former place, but though we heard of the arrival of the trains no boat had come. Arrangements had consequently again to be made for canoes & Indians, but we had scarcely turned in, when we heard a great row in the Indian village, with shouts of *boat chaco*, & sure enough before long we heard the sound of the oars in the rowlocks. On Saturday morning we got away in good time, & though experiencing some head wind, what with rowing, poling along shore & sailing when the wind was fair, we reached Fort St James at the S.E. end of the lake, as already stated late in the night. When we got near the camp, we roused them by firing a few shots, & soon had our stuff all piled on the bank, & tents in process of erection while we had a cup of tea at the camp fire. McConnell I found here, looking as if the trip agreed with him, & with all the things he started with from Victoria in good order.

From this place we hope to reach Fort McLeod in 5 or 6 days by the trail, when with the exception of self & McC. all will go down the Parsnip & Peace River in canoes or boats to Dunvegan. I go through the Pine River Pass to the same destination, with the whole of the pack trains & riding animals numbering in all, horses & mules, over one hundred. The boat parties will probably reach Dunvegan first, but we will make as good time as we can. This is the last Chance I know of for sending out letters for a time indefinitely long, so that no surprise need be felt if you do not hear from me for months. At the same time if any opportunity offers I will of course embrace it.

Having now written to you of past & future operations & plans at some length I must devote myself to other matters & correspondence, closing by sending best love to all. Please give benefits of contents of this letter to others to whom I shall not now write <at> in detail.

Yours very affectionately

George Dawson to Anna Harrington, Fort Stevenson, North Dakota, 17 May 1881.

My dear Anna,

As we are supposed to reach Ft Stevenson this evening, the first place with a name above Bismarck, I set down to write few lines to let you know how I am getting on. We left Bismarck very early yesterday (Monday) morning the whole of the precedent day having been spent in putting cargo on board. Yesterday afternoon we stuck on Mud bank no.1. in the midst of a shower of cold rain with heavy wind, & did not get off till after dark, when we tied up for the night. Today we have progressed without accident so far, but we must expect to be on many more before we reach Benton. Finding the steamer Red Cloud by which I had intended sailing, had to wait an indefinite number of days for the Police, who are coming up by steamer to Duluth, I took passage in the *C.K. Peck*. If we stick to some bank while the Red Cloud goes by it will be provoking enough, though I must say it takes a good deal to disturb my equanimity now one way or other. I seem to have lost all intensity of feeling of any kind or particular interest in anything. The baggage smashers succeeded in breaking my Camera box to pieces on the way to Bismarck, but I think the instrument itself can be so repaired as to be used, after reaching Benton. At St Paul I invested another dollar in light literature of the Cheap series, & have now gone through *Silas Marner*, *Sam Slick*, & *By the River* by Catherine Macquoid.[222] The scenery is monotonous, the company odious, & reading ones only resource.

Believe me

Your affectionate Brother

George Dawson to Anna Harrington, Fort MacLeod, North West Territories, (now Alberta), 15 August 1881.

My dear Anna,

Though you have proved my most constant correspondent, & I find no less than four of your letters in the bundle I have received since being in these parts, I fear I have been remiss in my duty of writing to you. I have just this afternoon returned to Macleod from a trip to the south & west, & have your kind letter of July 8th, from Metis, with others from all members of the scattered family save Eva, whose sickness accounts for her silence. I hope she is quite better long ago. In coming up the Missouri I made a few sketches, but without any marked success or pleasure, for I succeeded very badly in reproducing what I saw, & in accidentally turning up some of my old B. Columbia sketches in one of my books was disgusted to find more evidence of 'genius' in them than any I have done lately Sketching cannot be done mechanically or automatically, it requires some *elan*, & I dont think I will try any more till I feel this coming on, if I ever do. To begin with I am so busy now, from morning till night, & have so much riding & climbing & what not that my ideas have not time to focus themselves on sketching, or if accidentally concentrated form but a dim & cold point. Then when anything must be represented the Camera is always at hand. I am camped here on the brink of the Old Man River, about half a mile from Fort MacLeod, & have with determination begun this letter before turning in. The night is clear & almost cold, & starlight, outside, while the rippling of the river makes a pleasant sound as its clear blue water flows on toward the Saskatchewan, reminding me of "streams which swift or slow, draw down Aeonian hills & sow the dust of Continents to be" – or something of that sort. The horses are munching what grass they can find on the other side of the tent. Three of the men have gone into town to have a good time of it, while the Cook remains in Camp. McConnell, with the buck-board, a horse & man left this morning from Camp on the Waterton, eighteen miles from here, to go round by Kip to see some rocks there. He will be in here tomorrow night, probably.

During the last trip I had a rather pleasant expedition along the base of the Rocky Mts, northward. I had two men, two pack horses, & riding horses. We were gone a week, returning to Camp on Waterton Lake. Travelled chiefly through a partly wooded country with plenty brooks & small lakes, climbing over small spurs from the main range now & then. We had a small tent along but never pitched it, but spread our blankets just where it looked most comfortable on the grass or under the trees. The weather proved fine with the exception of one night, when rain pattering down wakened me. I got up & drew the waterproof sheet out from below the blankets & spread it on top, & so spent the remainder of the night comfortably enough. After getting back to the lake, when I met McConnell, we took a run up into

[222]George Eliot, *Silas Marner, the Weaver of Raveloe* (New York: Harper & Brothers, 1861); Thomas Chandler Haliburton, *The Clockmaker: Sayings and Doings of Samuel Slick of Slickville* (New York: Hurd and Houghton, 1874); and Katherine Sarah Macquoid, *Beside the River, a Tale* (London: Hurst and Blackett, 1881).

the mountains, chiefly for the purpose of procuring alpine plants, which are now just in perfection of spring beauty. Camped first night in a dense grove of spruce with a roaring fire, luxury unknown to the plains. Took horses to summit of pass next morning & then separated, McC going southward along the ridge, while I went northward. It was a blustry morning, but I got a couple of photos, which I hope may turn out well. Climbed up onto a peak about 1000 feet above the pass, from which a perfect panorama of snowy peaks in all directions. Wind blowing madly & seeming to go straight through one. Took a lot of bearings & made sketch of neighbouring mountains. Then came down again collecting plants en-route. The front of the mt. I was on ended eastward in a nearly sheer cliff of about 1000 feet or more, with a green valley with a little lake surrounded by patches of old snow drifts in the bottom, forming some of one branch of the brook in the pass. P.M. flurries of snow. Smoke of fire behind some scrubby pines, whirling excitedly in all directions, but chiefly into ones eyes. Squatted in most sheltered nook possible & proceeded to press plants for an hour or more.

About 5 Pm sheets of snow, whitening some of the higher mountain tops. Horses looked most miserable tied up to trees & scraping about among the stones for stray blades of grass As glad as ourselves to start back to Camp on lower level, where snow proved rain – after 10 P.m. up at 5 this mg. good night.

Aug. 7.

Here I am you see still at McLeod, but with every prospect of making an early start away tomorrow morning. This I shall have very great pleasure in doing as nothing is so tiresome as waiting about places like this, arranging supplies, settling difficulties with men, swapping horses &c. &c. As to reading I have done really none this summer, scarcely opening a book. Have not perused a single word of Victor Hugo, which has remained among my baggage in store here. My tent now is littered with papers, another mail having just arrived but to tell you the truth I feel very little interest in the news, out here it seems of little consequence. I will bundle them up & take a look now & then at them as I travel, we generally stop two & a half hours at noon for lunch & to let the horses feed & I sometimes have a few spare minutes. I am going to write a few lines to Father in case through any chance this general letter of news happens to miscarry. I had a letter from R. with mail found on arriving here, but do not want to write to him now as I should have to tell him what I think about his pursuing his trip to England, which would not be pleasant. He seems to want some new sensation every summer at someones expense, he will probably manage to get two or three trips across the Atlantic out of Father <out> by the last move, at a time when they can be ill afforded. I am very sorry to see by papers just received that Willie Redpath is dead, it must surely have been very unexpected as he always seemed such a strong athletic young fellow. It must be a great sorrow to Mrs R, & other members of the family. Tell Eric[223] I brought a pair of moccasins for him from an Indian, but I fear his feet may have grown so large before I get back that I may have to give them to Edith,[224] in which case he will be able to see the beadwork on them much more Easily, & to greater advantage. I am going northward to Bow River country next, but will detail plans in letter to Father. Kind regards to Bernard. I have letters from Father & Mamma from Dalhousie dated July 27th, which is extraordinary time, for this region.

Your affectionate brother

George Dawson to J.W. Dawson, Alert Bay, B.C., 2 August 1885.[225]

My Dear Father,

[223]George Eric Harrington (1878-1895), the oldest of Anna's children, was chronically ill.

[224]Edith Laura Harrington (1879-1890) was another of Anna's children who did not have a long life.

[225]In 1882 George travelled in Europe, and in 1883 and 1884 he explored more of interior British Columbia and Alberta.

We arrived here yesterday evening quite late, & as this happens to be Sunday are making it an off day. This place is opposite the mouth of the Nimpkish River, on the northern part of Vancouver Island. Postal communication is rather infrequent a steamer making monthly trips. We Just missed it in coming in last night & so this letter may have to wait three weeks or so, but I write in the hope of some chance turning up sooner. Alert Bay contains an Indian Village, a Salmon Cannery & Ch. of England Mission only, so that it is not a very lively place at the best of times & the Cannery has just been shut down for the season, the only work now going on there being the boxing etc. of the Cans. Salmon is so low priced at present that few of the Canneries along the Coast are doing much & a number are Closed altogether. Weather Continues fine & work progresses fairly, though not so fast as I could wish, & with rather monotonous results so far as rocks are concerned. Granitic rocks everywhere, with so far but one formation besides the Cretaceous, & this largely volcanic in an altered state & probably identical with the volcanic Triassic rocks of the Queen Charlotte Islands. I forgot to say that we found no mail here & in fact expected to find none as the notification of change of address from Comox did not reach Victoria in time to let the letters come up on the last steamer. I may be here, however, yet for a few days & shall be back later again when I hope to find a good collection of mail matter waiting. It will then be in order also to write again & at greater length with some chance of sending letter away.

Dowling,[226] one of your men whom you know is with me turns out very well. Is easy to get along with, quick, & is becoming quite a nautical character.

With kind regards to all the family – all I suppose enjoying themselves at Metis which is only another shore of this same great sea

Yours

**George Dawson to Anna Harrington, Alert Bay, B.C.,
6 September 1885.**

My Dear Anna,

Somehow I always forget to whom I wrote last, but I think it must now be some time since I wrote to you. We got back here yesterday, having crept round the north end of the Island from Quatsino through a series of Calms & fogs quite provokingly continuous. To Quatsino[227] I went round the coast in a canoe with two Nawitti Indians[228] & one of the men while the schooner followed. The Canoe voyage was an interesting one with a spice of adventure in it. The weather was very fine on the whole so that we had nothing to contend with except the everlasting long swell from the Pacific. This however made it possible to land only in sheltered spots in lee of points, behind islands or among reefs which broke up the sea. The Indians were very expert & I went round the whole coast just outside the surf, shooting through between rocks sometimes on the top of a wave & dodging in & out among the breaking water of reefs & shoals. There are two Indian villages in Quatsino Sound with a very unsofisticated lot of natives, the women having the singular & unique custom of

[226]Donaldson Bogart Dowling (1858-1925) had entered the service of the Geological Survey of Canada in 1884, and spent his entire working life with the survey, later leading many important western Canadian explorations.

[227]Quatsino Sound on the west coast of Vancouver Island, where Dawson had earlier visited in 1878.

[228]Kwakiutl Indians from Nawitti village on Hope Island.

elongating the head, so that they present a most remarkable appearance.[229] Something like this. [illustration in letter] The men keep the head as nature maked it. One white man, named Bowen, is living in the sound at present in charge of horses & Machinery used in boring for coal. He was very glad to see us as he had not seen any whites for three Months & found his post of guardian rather a lonely one. From here I propose making a tour round the north side of Queen Charlotte Sound as far as a point opposite the end of Vancouver Island, & shall then be working gradually southward & hope to pull up at Victoria before or about the end of next month. Should the weather hold fine into November I may be tempted to keep the field, or rather the sea, a little longer, but I think this hardly likely. This schooner is quite a weight of responsibility & I must say I shall be glad when I get her safely back to Victoria. I feel I should be on board when she is under way & have often to stay up most of the night on deck to see that all goes right & lay off courses etc by the Charts. When I get her safely moored in some harbour & go off in the boat examining the shores it is quite a holiday. Here, at Alert Bay, is one of the Centres of civilization & I hope to return & connect with mail in two or three weeks. There is here a salmon cannery with little work, however, going on at present, two or three whites & a few Chinamen only being employed putting up cases. – An Indian village consisting of the usual long row of low houses fronting the beach, & a church of England mission, with a small church & the mission house at the opposite end of the village from the Cannery. The Missionary, (A. Mr Hall)[230] & his wife have rather a hard time of it I think as the Indian is discouragingly obstinate & refuses to see the error of his ways. This note may have to lie here for the regular steamer – perhaps three weeks, – but I write hoping some chance may possibly occur before of getting letters down to Victoria.

I wish while the weather continues fine & after your return from the sea you would take a little trip somewhere for sketching & a general good time. If you can arrange to do so & find pleasant Company I wish to bear the expense of the expedition. I have not any way of sending money from here & my savings bank balance is nearly out, but I asked the people at Ottawa to forward my cheques for salary to father or William – I think to father – As father has my power of Attorney he can draw the needed amt. Please therefore ask him for say $50.00 & let me know about your expedition when I return.

Love to all. My excuses to Florence for having so far omitted to write to her. I suppose however she gets the benefit of my letters to others.

Yours

George Dawson to Anna Harrington, Near Glenora, B.C., 24 May 1887.

My dear Anna,

I write to you to report progress, which so far I regret to say is unsatisfactory. This is our fifth day from the sea in the little river steamer which runs on the Stickeen,[231] & with good luck, which we have no reason to expect, we may reach our destination at Telegraph Cr tonight. In that case the steamer will probably return tomorrow, taking down this letter. The river is a very swift one & difficult to climb, full of little rapids & in some places still shallow, as the summer rise has not yet fully commenced & our steamer has scarcely sufficient power. She is a little stern-wheeler without any accomodation except a few bunks for the hands & the rule therefore is to go ashore

[229]The Quatsino and neighbouring Koskimo practised this type of artificial head deformation which led to their being called "roundheads," "longheads," or "sugarloafs."

[230]Alfred James Hall (1853?-1918) was a missionary under the auspices of the Anglican Church Missionary Society who had moved to Alert Bay in 1880. For some thirty years Hall laboured there, establishing a residential school and sawmill.

[231]Stikine River.

& camp at night, about 8 o'clock, starting again between 3 & 4 in the morning. The steamer herself is quite a curiosity & till last night we had a cargo of about 40 chinamen, besides a few miners not chinamen. The celestials were, however, landed last night at Glenora, where there once was a lively little town, which, since the Cassiar Mines have gone down is almost completely abandoned. We have been for some hours at this spot endeavouring to get a line brought aboard to warp the steamer up a little rapid, having already tried unsuccessfully to steam up it. The little boat does the best she can with a tremendous pressure of steam on, far more than allowed by regulations or tests, but cannot wriggle up some of the swift water without a rope. From Telegraph we have 73 miles to go by pack train to Dease Lake, where my work is really to begin. From what I hear I doubt if the ice will have gone from the lake when we arrive there. The scenery on the Stickeen is very fine, as the river makes its way completely across the Coast Range to the sea, & is bordered by gigantic mountains. We are in a somewhat lower country now, in which the climate is relatively dry & the snow line is far up the mountains & trees leafing out. Lower down in the heart of the mountains there is a very heavy snowfall & the flats along the river are still more or less covered with snow. There are also several very large glaciers, one of which pushes out almost to the river, with a front two miles or so in length. The weather is very fine & you may suppose I am very anxious to get on to the lake

Yours George

I fear there is likely to be a delay of some days here, & if so will write again to someone at home. Meanwhile the steamer is to leave again in a few hours & I take the chance of sending this down to Wrangell, where it may chance to get off or remain for a week or two before leaving for Victoria

Yours
George

25 May

P.S. Since writing above have arrived at Telegraph Creek, the head of navigation on the Stickeen. It took us all day yesterday & today till noon to make the distance from Glenora, about 12 miles. We came in fact near loosing the steamer altogether in one of the rapids, as she took a shear & was carried back against a high rock, from which, however, she rebounded comparitively uninjured. The captain says it is the roughest trip he has made on the river so that we may count we have had a pretty good specimen of the navigation.

George Dawson to Margaret Dawson, Revelstoke, B.C., 10 June 1889.

My dear Mother,

I write just a line to let you know that I have arrived here & am to leave tomorrow to go down the Columbia & then in to Kootanie Lake. I have got a good boat & I think two good men, besides Mr Edgar & I hope we may make a fairly enjoyable trip of a month or so, when I shall hope to return to this place & again strike the railway. The weather is now very fine, but uncommonly warm, quite melting & up in the ninetys today, but always cool & pleasant at night. This is a horrible shabby little village which was a bustling place during railway construction, but has since collapsed & now only shows signs of renewed vitality in consequence of mining developments occurring in this vicinity. Fine snow covered mountains all around & an expanse of burnt stumps everywhere in the foreground. As I have really nothing more to say than what is above stated & as I left Kamloops by train at 5 this morning I think I shall now do well to turn in.

I hope Father has been able to see his way to make some arrangement to relieve him of part of his work at least, & I hope also that you will remember that you may count on me to afford any assistance in my power toward this end. I have of course not yet had time to hear from home & shall now scarcely receive any news till I get back from Kootanie in about a months time

Yours

George Dawson to Margaret Dawson, Ashcroft, B.C., 31 August 1889.

My dear Mother,

I got here yesterday evening & we have been devoting the day to refitting & general overhaul preparatory to another tour. I had mail sent down from Kamloops & found amongst other letters one from Father in which he spoke of shortly going to Toronto (date Aug. 19th) I trust he may have a pleasant time there. My work seems never to come to an end & rather grows as I go on with it, but I hope to have the greater part of the task outlined in the spring, overcome before leaving the field. Much must of course depend on the weather, but under ordinarily favourable circumstances we may hope for six weeks of good working time yet, & it is probable that I shall be back in the east not long after November first. We have here one of our not very enjoyable settlement camps, on the opposite side of the Thompson from Ashcroft station, with the rushing river in front & steep high clay banks behind. {Fortunately} there is a good bridge not far off by which we reach the 'town'. Both the east & west mail trains pass here in the night, so that it is a place no one ever sees, but has some importance as being now the point for which traffic starts by road for the 'Upper Country' – Cariboo etc. The very hot weather seems now to have come to an end, & even here in the low Thompson Valley the nights are cool. During the first week we have had much cool windy & broken weather & as we were part of the time on high mountains – camped a couple of nights at 6800 feet above sea – we came in for full benefit. Squalls of snow all one day at frequent intervals. Salmon are very plentiful in the Fraser & its tributaries this summer. The canneries at mouth of river have made large packs, & everywhere along the Fraser & Thompson the Indians have gathered & dried great quantities of fish. Just at our camp is a staging of poles with hundreds of dried salmon. The Indians have in fact now given up catching the fish & appear to be satisfied that they have an ample supply for the winter. I write this sitting at the fire to the accompayment of the rushing sound of the river & the chirping of innumerable crickets on the hill behind. The men are all off to the town on various errands. It is getting so dark that I find difficulty in following the lines any longer, as I daresay you may have noticed before this, & so will conclude by sending love to all at home.

Yours

George Dawson to Anna Harrington, Kamloops, B.C., 13 July 1894.[232]

My dear Anna,

After a cruize through the foot-hills between Macleod & Calgary, with Mr McConnell, I have got across here ready for my more serious work for the summer. McEvoy, who is to assist me here has gone down to Victoria to bring up some of our stuff which had been stored there. He cannot get back till tomorrow (Saturday) night, after which comes Sunday so that I am stalled here as it were for a few days, without very much to do except looking after repairs of 'rigging' etc. All the stuff we left here was in a storehouse which was submerged under six feet of water during the flood & as you may imagine the saddles, kitchen &c &c are in a rather poor state in consequence, though now dry. We had a pretty hot time in the foot-hills, which with high torrid winds brought about a complete exfoliation of my face, which it is now only beginning to recover from. Here it is also hot – over 90° in the shade today, but dry & always beautifully cool & airy in the evenings, while for beautiful colouring of the hills & mountains after sunset there is no place that I know like Kamloops. The evening is a special enjoyment every day.

I hope that matters have been prospering with you since you got to the Adarondacks & that you have found your quarters comfortable, or at least have been able to find suitable & comfortable quarters

[232]In the intervening years George spent most summers in the Kamloops region in the interior of British Columbia, refining his geological concepts.

somewhere. I hope also that the change has been beneficial to Eric. I did not stop at Morley on the way as I could not get a train from Calgary there which would enable me to make the trip by daylight, which I was anxious to do to see again some features of the Bow Valley. I ascertained, however, that D. McDougall is in the habit of taking in strangers as there is no regular hotel there. His place is clean & might be interesting for a short time. Very pretty scenery & fishing in the Bow R. near bye. Then there is Banff & further on this place. Here the hotel accomodation is somewhat primitive, but clean & comfortable enough. The resources are not very great, except perhaps in the way of riding over the bunch-grass hills – now already yellow & sore looking. There is a local band which plays some evenings in the week & the usual going & coming of a small town. Several churches of different Kinds. As I write the frogs are croaking along the edge of the river – not cherriping like the eastern frogs, but regularly croaking & a very hoarse croak at that. The perennial river is bordered by very green cottonwood trees forming the greatest contrast possible to the general dryness.

Should you find it advisable to come west I hope you will continue to remember that there need be no difficulty whatever in regard to the cost of the trip. You have only to let me know.

Yours

George Dawson to Anna Harrington, Savona, B.C., 5 August 1894.

My dear Anna,

Yours of July 21st reached me here today & I am glad to know of your continued comfort & also to hear of the slight improvement you note in Eric. Bernard's return with the continued good accounts which come from Metis must also relieve your mind of some responsibility. It is fortunate that you have been able to make up your mind not to worry about how matters there are going & I think that your stay in the Adirondacks will be a useful rest to you personally if you can continue to take it in that feeling. You must have plenty of time & I think that it should be pleasant for you to renew your sketching habits where there must be so many pretty places not far off. I have just got in here from the hills, having completed one branch of my work i.e. that connected with revision of part of the 'Kamloops sheet'.[233] This has meant riding everlastingly along dusty tracks & up & down steep stony hills. Travelling sometimes twenty miles in the course of the day & often as much as eight hours in the saddle with the thermometer ranging anywhere above 90°. This sounds like hard work, but one very soon becomes used to it & it gives you a fair appetite & a very satisfactory thirst with good sleep & all that sort of thing. We have only had one mountain camp so far at 6000 feet up & there it was bitterly cold for a few hours, in fact a few pellets of hail fell. I am now about to start for the Horsefly Country in Cariboo direction where I hope to spend a few days looking into the hydraulic mining begining there. This will be partly a stage trip & I am not taking any camping outfit but trusting to the accomodation of 'hotels' by the way where one is much less comfortable than in camp with poorer food comprising unwholesome pies & washy tea. When I get back, which may be in ten days or so, I hope to spend a little time canoeing on some of the lakes, which should be pleasant as a change & implies comparatively little exertion.

My best wishes to Eric & Believe me

Yours

[233]George M. Dawson, *Kamloops Sheet, British Columbia, Geologically Coloured* (Ottawa: Geological Survey of Canada, 1895).

PIONEERING WORK IN A NEW LAND: DAWSON AND THE YUKON

George Dawson spent so many years of his life in geological exploration and research that in later years he was able to look back with profound satisfaction, knowing that his observations and reasoning on geological matters and his anticipation of mineral wealth in specific regions had been born out by later events. The development of the Kootenay region, the hydraulic mines of the Cariboo, and, especially, the Yukon gold mines he all foretold in his earlier reports.

Dawson's geological explorations of Canada led him into virtually unknown country and were carried out under conditions which would have taxed the endurance of many stronger men: for instance, his exploration of the Yukon and adjacent portions of northern British Columbia. He always had a special interest in these great northern regions about which so little was known. In his voluminous and much sought after reports upon the resources of the areas he examined, will be found the most authentic and useful information on those now flourishing districts.

In 1887, he commanded the Yukon expedition, a remarkable thirteen hundred mile boat journey of zeal and energy. Later, after a careful search through accounts left by northern explorers and examination of geological collections brought back by some, Dawson published a geological map of the northern portion of the Dominion of Canada east of the Rocky Mountains with accompanying notes, in which all existing information about the geology of this remote region was set forth.[234]

It was said later that:

> *In his office in the Canadian Geological Survey in Ottawa, George Dawson unrolled his maps, and was able to show the geological contour of any section into which the miners might choose to go. Within a few days of the news that gold had been found in the Klondike, the newspapers were able to give a description of the region and a glowing account of what might be expected from it, all based upon the work Dawson had done – Yet with all this knowledge in his keeping, he guarded it carefully from any unscrupulous use; had he been a self-seeking and dishonest man he might have, through his knowledge, added immeasurably to his small salary of $3,000.00.*

Some years after Dawson's death, R.W. Shannon of Ottawa wrote:

> *Klondike Dawson was a personage in his own right at the Rideau Club in Ottawa – and he was a gay companion, physical weakness never depressed his bright spirits – his constant cheerfulness was a suprise to all who knew him, more especially those who reflected upon the fortitude required to bear his bodily infirmities with patience. In conversation, he was witty and humor-*

[234]George M. Dawson, "Geological Map of the Northern Part of the Dominion of Canada, East of the Rocky Mountains. . .," in Geological Survey of Canada, *Annual Report 1887*, n.s., 3 (1889), Vol. II, Part R.

ous to a degree, while at the Club, or at a public dinner, his sallies were wont to keep the table in roars of laughter – but over the sparkling glass and silver of the Rideau club or while at work at his desk at the office of the Survey, his mind would oft times go back to days of travel into the unknown Yukon among the 'hard-fists' of the wilderness to whom he had been a good companion.

One by the name of Davison also said, "It was at night round the Camp fires that he opened up, it was a treat to listen to him."

Dawson's deep interest in a wide range of Yukon subjects is revealed in the following excerpts. Also, even when written for popular consumption, Dawson's articles contain accurate and detailed information, founded on solid reasoning.

Historical Notes on the Yukon District (Toronto: William Briggs, 1898), 16-17.

In 1896, D. W. Davis was appointed collector of customs for the Yukon district. During the early summer most of the miners were employed on the branches of Forty-mile and Sixty-mile Creeks, but about 100 men were reported to be working along the Teslin or Hootalinqua. Late in August, "coarse" gold was discovered by G. W. Cormack in the Klondyke valley. The richness of the find became established before the end of the year, and a "rush" occurred. Forty-mile and Sixty-mile Creeks were nearly abandoned and the population of Circle City, Alaska, (more than 100 miles below the boundary) was reduced from about 1,000 to about 300. Dawson, or "Dawson City" was laid out by J. Laduc at the mouth of Klondyke Creek. Glacier and Miller Creeks had been, up to this time, the richest discovered. Early in the summer, Mr. J. E. Spurr of the U.S. Geological Survey, with two assistants, crossed by the Chilkoot pass and descended the river for the purpose of exploring that part of the gold-bearing region which extends into Alaska. Forty head of cattle were this summer driven in over the "Dalton trail" from Chilkat to Fort Selkirk. Dalton, by whose name the trail is known, had already crossed several times by this route, from 1894 or perhaps even earlier, but had not made it generally known. The arrival of deserters from the whaling vessels at Herschel Island overland *viâ* Rampart House on the Porcupine, is mentioned in the police report as having occurred annually for some years. The value of gold obtained in the Yukon district in 1896 is estimated at $300,000.

In the spring of 1897, T. Fawcett was sent to the Yukon district as gold commissioner, with two assistant surveyors. Twenty-five police with an officer were also despatched *viâ* the Chilkoot pass to relieve those in the country who had engaged for two years only. J. A. McArthur and A. St. Cyr were sent by the surveyor-general to examine the Chilkat pass and Dalton trail, and the country between the Stikine and Teslin Lake, respectively. The continued great influx of population led, later in the season, to the appointment, from August 15th, of Major J. M. Walsh as chief executive officer for the Yukon district. Judge McGuire of the Supreme Court of the North-west Territories was transferred to the Yukon and Mr. F. C. Wade was appointed registrar, crown prosecutor and clerk of court. Since September 1st, additional detachments of police, aggregating 100 men, have been despatched to the Yukon district. In August, Mr. T. W. Jenning, with assistants, was sent by the Canadian government to examine a route *viâ* the Stikine and overland from Telegraph Creek, the head of navigation on that river, to Teslin Lake, with a view to the construction of a railway. In October, the Hon. Clifford Sifton, Minister of the Interior, crossed the Chilkoot pass to the lakes and returned to the coast by the White pass, in order to ascertain the precise conditions prevailing there. An unprecedented rush of miners and others set in during the summer to the Yukon district, the majority going by way of the Chilkoot and White passes, from the head of Lynn Canal, some ascending the Yukon from its mouth on Behring Sea, and others filtering in by various channels. The results of this movement of population, both in the district and elsewhere, are still engaging the attention of the public and the press.

The value of gold produced in 1897 is roughly estimated at $2,500,000, an amount greater by half a million dollars than that obtained from the Cariboo district of British Columbia in 1861, the year of the discovery and first working of Williams and Lightning Creeks.

"Notes on the Occurrence of Mammoth-Remains in the Yukon District of Canada and in Alaska," *Quarterly Journal of the Geological Society* 1 (1894), 6-7.

In the present connexion, the 'ground-ice formation' is of interest only in so far as its existence and the mode of its origination may throw light on the date and method of entombment of the Mammoth-remains associated with it. With respect to the origin of the deposits, the writer ventures to offer the following suggestions.

The country in which the 'ground-ice formation' occurs is low in its relief, and the formation occupies its lower tracts. The ice itself must undoubtedly have been produced upon a land-surface, and since the time of its production this surface can never have been covered by the sea; for this would inevitably have reduced the frozen condition of the overlying clays, and have resulted in the destruction of the icy substratum as well.

With an elevation of the land by an amount of 300 feet or more (such as appears to be required by the Mammoth-remains on islands already mentioned) the warmer water connecting with the Pacific would be confined to the deeper western portion of what is now Bering Sea, forming there a limited gulf, without outlet to the north, from which the region where the 'ground-ice formation' is now found would be so far removed as to greatly reduce its mean annual temperature. Snow falling upon this nearly level, northern land, and only in part removed during the summer, would naturally tend to accumulate in *nevée*-like masses in the valleys and lower tracts, and the underlying layers of such accumulations would pass into the condition of ice, though without the necessary slope or head to produce moving glaciers. The evidence does not seem to imply that the Mammoth resorted to this extreme northern portion of the region during the actual time of ice-accumulation, but this animal may be supposed to have passed between Asia and America along the southern parts of the wide land-bridge then existing.

At a later date, when the land became depressed to about its present level, Bering Sea extended itself far to the eastward, and Bering Straits were opened. The perennial accumulation of snow upon the lowlands ceased, and in the southern parts of Alaska such masses as had been formed may have been entirely removed. Farther to the north and at a greater distance from the Pacific waters, while the total precipitation would probably be increased, a greater proportion would fall as rain, and floods resulting from this and the melting of snow on the higher tracts would be frequent. Thus it may be supposed that deposits of clay and soil from adjacent highlands and from the overflow of rivers covered large parts of the remaining ice of the lowlands, and that wherever so covered it has since remained; the winter temperature being still sufficiently low to ensure the persistence of a layer of frozen soil between the surface annually thawed and the subjacent ice. Over the new land thus formed the Mammoth and associated animals appear to have roamed, and fed, and wherever local areas of decay of the ice may have arisen, bottomless bogs and sink-holes must have been produced which served as veritable traps.

"Notes on the Indian Tribes of the Yukon District and Adjacent Northern Portion of British Columbia," in Geological Survey of Canada, *Annual Report 1887*, n.s., 3 (1889), Report B, Appendix II, 192-94.

The division of the Tinné met with on ascending the Stikine is named Tahl-tan, and consists of the Tahl-tan people proper and the Taku. These Indians speak a language very similar to that of the Al-tá-tin, if not nearly identical with it, and so far as I have been able to learn, might almost be regarded as forming an extension of the same division. They appear to be less closely allied by language to the Kaska, with which people they are contiguous to the eastward.

The Indian village near the Tahl-tan or First North Fork of the Stikine, is the chief place of the Tahl-tan Indians, and here they all meet at certain seasons for feasting, speech-making and similar purposes. The Tahl-tan claim the hunting-grounds as far down the Stikine, coastward, as the mouth of the Iskoot River, together with all the tributaries of the Iskoot and some of the northern sources of the Nass, which interlock with these. Their territory also includes, to the south, all the head-waters of the main Stikine, with parts of adjacent northern branches of the Nass. Eastward it embraces Dease Lake, and goes as far down the Dease River as Eagle Creek, extending also to the west branch of the Black or Turnagain River. It includes also all the northern tributaries of the Stikine, and the Tahl-tan River to its sources.

The Taku form a somewhat distinct branch of the Tahl-tan, though they speak the same dialect. They are evidently the people referred to by Dall [Contributions to North American Ethnology.] as the Tāh́-ko-tiń-neh. They claim the whole drainage-basin of the Taku River, together with the upper portions of the streams which flow northward to the Lewes; while on the east their hunting-grounds extend to the Upper Liard River, and include the valleys of the tributary streams which join that river from the westward. They are thus bounded to the south by the Tahl-tan, to the west by the coast Taku (Thlinkit), to the north-west by the Tagish, and to the east by the Kaska.

The territorial claims of the Tahl-tan and Stikine Coast Indians (Thlinkit) overlapped in a very remarkable manner, for while, as above stated, the former hunt down the Stikine valley as far as the Iskoot, and even beyond that point, the latter claimed the salmon-fishery and berry-gathering grounds on all the streams which enter the Stikine between Shēk's Creek (four miles below Glenora) and Telegraph Creek, excepting the First South Fork, where there is no fishery. Their claim did not include Telegraph Creek nor any part of the main river; nor did it extend to the Clearwater River or to any of the tributaries lower down. In whatever manner the claim to these streams may have been acquired, the actual importance of them to the Coast Indians lay in the fact that the arid climate found immediately to the east of the Coast Ranges enabled them to dry salmon and berries for winter provision, which is scarcely possible in the humid atmosphere of the coast region.

The strict ideas entertained by the Indians here with respect to territorial rights is evidenced by the fact that the Indians from the mouth of the Nass, who have been in the habit of late years of coming in summer to work in the gold mines near Dease Lake, though they may kill beaver for food, are obliged to make over the skins of these animals to the local Indians. Thus, while no objection is made to either whites or foreign Indians killing game while travelling, trapping or hunting for skins is resented. In 1880 or 1881, two white men went down the Liard River some distance to spend the winter in trapping, but were never again seen, and there is strong circumstantial evidence to show that they were murdered by the local Indians there.

On the Stikine, as in the case of other rivers and passes forming routes between the coast and the interior, the Coast tribes assumed the part of middle-men in trade, before the incursion of the miners broke up the old arrangements. The Stikine Indians allowed the Tahl-tan to trade only with them, receiving furs in exchange for goods obtained on the coast from the whites. The Tahl-tan, in turn, carried on a similar trade with the Kaska, their next neighbors inland. The right to trade with the Tahl-tan was, in fact, restricted by hereditary custom to two or three families of the Stikine Coast Indians.

With the exception of the houses already referred to as constituting the Tahl-tan village, and some others reported to exist on the Taku, the residences and camps of these people are of a very temporary character, consisting of brush shelters or wigwams, when an ordinary cotton tent is not employed. We noticed on the Tahl-tan River a couple of square brush houses formed of poles interlaced with leafy branches. These were used during the salmon-fishing season. At the same place there were several graves, consisting of wooden boxes or small dog-kennel-like erections of wood, and near them two or three wooden monumental posts, rudely shaped into ornamental (?) forms by means of an axe, and daubed with red ochre.

On attaining the chieftaincy of the Tahl-tan tribe, each chief assumes the traditonal name Na-nook, in the same manner in which the chief of the Coast Indians at the mouth of the Stikine is always named Shēk or Shake.

The Tahl-tan Indians know of the culture- or creation-hero Us-tas, and relate tales concerning this mythical individual resembling those found among the Tinné tribes further south, but I was unable to commit any of these to writing. Amongst many other superstitions, they have one referring to a wild man of gigantic stature and supernat-ural powers, who is now and then to be found roaming about in the summer season. He is supposed to haunt specially the vicinity of the Iskoot River, and the Indians are much afraid of meeting him.

ETHNOLOGICAL ENDEAVOURS

George Dawson's ethnological achievements were admirably summarized by W.J. McGee, who commented that:

> *While several of Dr Dawson's titles and the prefatory remarks in some of his papers imply that his ethnologic researches were subsidiary to his geologic work, and while his busy life never afforded opportunity for monographic treatment of Canada's aborigines, it is nevertheless true that he made original observations and records of standard value, that much of his work is still unique, and that his contributions, both personal and indirect, materially enlarged knowledge of our native tribes. It is well within bounds to say that, in addition to his other gifts to knowledge, George M. Dawson was one of Canada's foremost contributors to ethnology, and one of that handful of original observers whose work affords the foundations for scientific knowledge of the North American natives.*[235]

Dawson's most noteworthy ethnological work was undoubtedly his observations on the Haida Indians of the Queen Charlotte Islands,[236] *but he also published notes on the Indian Tribes of the Yukon District and adjacent Northern Portion of British Columbia,*[237] *as well as valuable memoirs entitled 'Notes on the Kwakiool People of Vancouver Island' and 'Notes on the Shuswap People of B.C.' In association with W. Fraser Tolmie he prepared a valuable 'Comparative Vocabularies of the Indian Tribes of B.C., with a map illustrating distribution.'*[238] *When, in 1884, the British Association for the Advancement of Science appointed a Committee on the North-western Tribes, to study the physical characteristics, languages, and social conditions of the Northwest Coast Indians, Dawson was made a member and arranged funding so that the many artifacts collected for the committee by the prominent anthropologist Franz Boas would stay in Canada. In fact, Dawson later took effective direction of the committee and, in 1897, chaired its successor, the Ethnological Survey of Canada.*

[235] McGee, "George Mercer Dawson," 162.

[236] "On the Haida Indians of the Queen Charlotte Islands," Geological Survey of Canada, *Report of Progress for 1878-79* (1880), Report B, Appendix A.

[237] "Indian Tribes of the Yukon."

[238] "Notes and Observations on the Kwakiool People of the Northern Part of Vancouver Island and Adjacent Coast. . . ," *Proceedings & Transactions of the Royal Society of Canada* 5 (1887), Sec.II, 63-98; "Notes on the Shuswap People of British Columbia," *Proceedings & Transactions of the Royal Society of Canada* 9 (1891), Sec.II, 3-44; and W. Fraser Tolmie and G.M. Dawson, *Comparative Vocabularies of the Indian Tribes of British Columbia, with a Map Illustrating Distribution* (Montreal: Dawson Brothers, 1884).

The following are a few remarks about the Haida Indians of the Queen Charlotte Islands, taken from George Dawson's articles on that subject.

"The Haidas," *Harper's New Monthly Magazine* 45 (August 1882), 404-5.

In their mode of life, and the ingenuity and skill they display in their manufacture of canoes and other articles, the Haidas do not differ essentially from the other tribes inhabiting the northern part of the coast of British Columbia and Alaska. In the Queen Charlotte Islands, however, the peculiar style of architecture and art elsewhere among the Indians of the west coast more or less prominently exhibited, appears to attain its greatest development. Whether this may show that to the Haidas or their ancestors the introduction of this is due, or indicate merely that with the greater isolation of these people, and consequent increased measure of security, the particular ideas of the Indian mind were able to body themselves forth more fully, we may never know. The situation of the islands, and the comparative infrequency with which they have been visited for many years, have at least tended to preserve intact many features which have already vanished from the customs and manufactures of most other tribes.

...The general type of construction of houses with the Indians of this part of the northwest coast is everywhere nearly the same, but among the Haidas they are more substantially framed, and much more care is given to the fitting together and ornamentation of the edifice than is elsewhere seen. The houses are rectangular, and sometimes over forty feet in length of side. The walls are formed of planks split by means of wedges from cedar logs, and often of great size. The roof is composed of similar split planks or bark, and slopes down at each side, the gable end of the house–if such an expression may be allowed–facing the sea, toward which the door also opens.

The door is usually an oval hole cut in the base of the grotesquely covered post, forty or fifty feet high, which we may call the totem post, but which to the Haidas is known as *kechen*. Stooping to enter, one finds that the soil has been excavated in the interior of the house so as to make the actual floor six or eight feet lower than the surface outside. You descend to it by a few rough steps, and on looking about observe that one or two large steps run round all four sides of the house. These are faced with cedar planks of great size, which have been hewn out, and serve not only as shelves on which to store all the household goods, but as beds and seats if need be. In the centre of a square area of bare earth the fire burns, and it will be remarkable if some one of the occupants of the house be not engaged in culinary operations thereat. The smoke mounting upward passes away by what we may call a skylight–an opening in the roof, with a shutter to set against the wind, and which serves also as a means of lighting the interior. One is surprised to find what large beams have been employed in framing the house. There are generally four of these laid horizontally, with stout supporting uprights at the ends. They are neatly hewn, and of a symmetrical cylindrical form, and are generally fitted into the hollowed ends of the uprights. The uprights are often about fifteen feet high, with a diameter of about three feet; and it is only when we become acquainted with the fact that a regular *bee* is held at the erection of the house that we can account for the movement without machinery of such large logs. The bee is accompanied by a distribution of property on the part of the man for whom the house is being built, well known on the west coast by the Chimook name *potlatch*. Such a house as this accommodates several families, in one sense of the term, each occupying a certain corner or portion of the interior.

"On the Haida," 148.

The peculiar carved pillars which have been generally referred to as carved posts are broadly divided into two classes, known as kexen and *xat*. One of the former stands at the front of every house, and through the base, in most instances, the oval hole serving as a door passes. The latter are posts erected in memory of the dead.

The *kexen* are generally from 30 to 50 feet in height, with a width of three feet or more at the base, and tapering slightly upwards. They are hollowed behind in the manner of a trough, to make them light enough to be set and maintained in place without much difficulty. These posts are generally covered with grotesque figures, closely grouped together, from base to summit. They include the totem of the owner, and a striking similarity is often apparent between the posts of a single village. I am unable to give the precise signification of the carving of the posts, if indeed it has any such, and the forms are illustrated better by the plates [in report] than by any description. Human figures, wearing hats of which the crowns run up in a cylinderical form, and are marked round with constrictions at intervals, almost always occur, and either one such figure, or two or three frequently surmount the end of the post. Comparatively little variation from the general type is allowed in the *kexen*, while in those posts erected in memory of the dead, and all I believe called *xat*, much greater diversity of design obtains. These posts are generally in the villages, standing on the narrow border of land between the houses and the beach, but in no determinate relation to the buildings. A common form consists of a stout, plain, upright post, round in section, and generally tapering slightly downwards, with one side of the top flattened and a broad sign-board-like square of hewn cedar planks affixed to it. This may be painted, decorated with some raised design, or to it may be affixed one of the much prized 'coppers' which has belonged to the deceased. In other cases the upright post is carved more or less elaborately. Another form consists of a round, upright post with a carved eagle at the summit. Still others, carved only at the base, run up into a long round post with incised rings at regular intervals. Two round posts are occasionally planted near together, with a large horizontal painted slab between them, or a massive beam, which appears in some instances to be excavated to hold the body. These memorial posts are generally less in height than the door posts.

The paragraphs that follow are taken from Dawson's article, "Sketches of the Past and Present Condition of the Indians of Canada," Canadian Naturalist & Quarterly Journal of Science *9 (1881), 129-30, 133, 135-36, 157-58, which also had been read in Edinburgh.*

Constituting thus nearly a fortieth part of the entire population of Canada, the Indians would even numerically be a not unimportant factor in questions of interior policy. As the original possessors of the land, however, though possessing it in a manner incompatible with the requirements of modern civilization, and as having been at times ready to assert that ownership, even in a forcible manner, they acquire quite a special interest; even without that afterglow of romance which follows the memory of the red man in those regions from which he has already passed away.

Though in the ante-Columbian period of American history nearly all the Indian tribes and nations appear to have been either drifting or gradually extending, by force of arms, in one direction or another, as indicated by their history or traditions, their movements were neither so rapid nor erratic as those which have occurred since the old organization and balance of power began to crumble before the advance of irresistible force from without. We may therefore trace, with some degree of definiteness, the extension of the greater Indian families as they existed when first discovered, grouping together, for this purpose, many tribes which, though speaking the same or cognate languages, and with a general similarity in habits and modes of life, were not infrequently at bitter enmity among themselves, and in some cases had almost forgotten their original organic connection.

...The intercourse of the Europeans and Indians of the north-eastern portion of America can scarcely be said to have been begun by Cabot in his voyages of 1497-98-99, when he first discovered this part of the coast. With Cartier, in 1534 and 1535, in his memorable voyages up the St. Lawrence, the first real contact occurred. The natives appear to have received him often timidly, but were found ready enough to trade when friendship had been cautiously established. At the villages of Stadacona (Quebec) and Hochelaga he was received even with rejoicing, the natives bringing gifts of fish, corn and "great gourds," which they threw into his boat in token of welcome. It is evident, however, that they well understood and wished to maintain their territorial rights, for we find that when Cartier, in his first voyage, set up in the vicinity of the Baie des Chaleurs his "cross thirty feet high," the aged chief of the region objected to the proceeding, telling the French–as well as his language could be understood–that the country all belonged to him, and that only with his permission could they rightly erect the cross there. It was too, when, in 1541, Cartier attempted his abortive colony at Quebec, that the natives first manifested jealousy and a hostile spirit.

...We may well ask upon what principle they have been remunerated for their lands; certainly not by any standard either of their absolute or relative value, rather by that of the relative ignorance of the various tribes at the time they were treated with, and the urgency of their then present wants. Looked at from this point of view, the transaction loses altogether the aspect of an equitable purchase. It must be evident that the Government, in such arrangements, does *not* fully acknowledge the Indian title, the "territorial estate and eminent dominion" being vested in the crown, and the claim of the Indians restricted practically–though not patently in the transactions as effected with the Indians–to right of compensation for the occupancy of their hunting grounds.

...It is often said that the ultimate fate of the Red Man of North America is absorption and extinction : just as European animals introduced into Australia and other regions, frequently drive those native of the country from their haunts, and may even exterminate them, and as European wild plants accidentally imported, have become the most sturdy and strong in our North American pastures; so the Indians races seem to diminish and melt away in contact with the civilization of Europe, developed during centuries of conflict in which they have had no part, but during which their history has moved in a smaller circle, ever returning into itself. Even the diseases engendered in the process of civilization, and looked upon in the Eastern hemisphere with comparative indifference, become, when imparted to these primitive peoples, the most deadly plagues. Dr. J. C. Nott (as quoted by Prof. Wilson), writes: "Sixteen millions of aborigines in North America have dwindled down to two millions since the Mayflower discharged on Plymouth Rock; and their congeners the Caribs have long been extinct in the West Indian Islands. The mortal destiny of the whole American group is already perceived to be running out, like the sand in Time's hour-glass." Dr. Wilson, has, however, himself shown that though the Indian as such can not very much longer survive, Indian blood in quantity quite inappreciated by casual observers now courses through the veins of white persons of the continent.

The ultimate object of all Indian legislation must be, while affording all necessary protection and encouragement during the dangerous period of first contact with the whites, to raise the native eventually to the position of a citizen, requiring neither special laws of restraint or favour. When it is found that the paternal care of the State begins to act as a drag on the progress of the Indian, and that after reaching a certain stage all further advance ceases, the state of dependence must be done away with.

GLIMPSES AT GEORGE'S PERSONAL LIFE

My uncle's spirits never appeared high when he had to be in cities for long, and it would seem that his spirits were at an especially low ebb in 1881 and 1882. According to my mother, there had been a breaking off of a very personal relationship with one named Emma, after which Dawson's mood became dull and his zest for life weakened. Apparently, he was very much in love, but marriage was not permitted by her parents.

The following poems no doubt relate to this time.

"My Love if thou dost hold the wine of two Men's
lives in thy dear hands,
I pity thee, for thou hast what thou can'st not restore
If thou bearest one away in thy sweet heart
Then must thou spill the other in the sand.
But blame me not if I do pray thee for my soul
Oh leave me not to mourn the empty cruise
the evening of my days
Long time in secret has the fragrance grown
It is my all I pray thee for my own."

"You are that note from early dawn
That sounds through life however long,
The pristine music of the race
We can but name the morning song
The world is old, and I am old
Gray hairs grow thick, some honours fall
But that one day when you and I were one
is still the best of all
So now come death, or chance what may
In downward slope of passing years
I hold the memory of a day."

Much of 1882, from 5 May to 4 December, George spent in Europe, when he made a tour of mines and metallurgical works. His emotional state, then, was obviously fragile and depression coloured his world view, as evidenced by his letters to Anna.

George Dawson to Anna Harrington, Nevers, France, 20 June 1882.

My dear Anna,

I left Paris this morning a little before noon, & arrived here at 7.30 why I am here it would hard to explain. I am supposed to be on my way to Cleremont & the Auvergne country on a geological excursion, but thinking it would be rather far to go to Clermont in one day looked this place up on the map & decided to stay here all night. It is a stupid & dull little place & at present perfectly saturated, as it has been raining all day. I got perfectly disgusted with Paris seeing the same streets & same advertisements every day & so decided to leave. I wish some one who could derive pleasure or profit from travelling was here instead of me. If I could only sleep all day as well as at night I would be moderately happy. If Father were here in France he would be running about with the greatest interest getting collections & seeing what was to be seen. I seem to be too old to take any pleasure in such things; & geology interests me, well fully as much as astrology for instance. Yesterday night I was at the Grand opera in Paris & saw a piece called Aida. The music is by Verdi & I believe the piece is new, or at least comparitively so. The scene is in Egypt & much care has evidently been taken to reconstruct the surroundings as they should be. The dresses &c of course magnificent. The opera receives about 400,00 francs annually from govt. besides what they take for their 2000 seats & so is able to do things in style. At one time there were I estimated, about 400 people on the stage forming a crowd of soldiers, priests priestesses &c. &c.

The hay is all cut in this part of the country & the barley & oats beginning to turn yellow, so that the harvest cannot be far off. Much of the ground hereabouts is taken up with vineyards in which the vinegar known as vin ordinaire & forming the beverage of the people is produced. If the grapes were not so poor one would be inclined to think them wasted in being converted into such wine.

Travelling *en garcon* is not a very lively occupation at best & I had such a hideous old french woman opposite me in the railway today that I have not yet recovered from the mental impression received, & moreover fear nightmares.

I suppose you will be at Metis by the time this reaches, & so address it there. Hope you may have a pleasant summer.

Excuse more at present

Yours

George Dawson to Anna Harrington, Lyons, France, 7 July 1882.

My dear Anna,

I received quite a budget of letters here, including those of two weeks. I cannot answer them all, & besides have no particular news to communicate, however, as 'my plan' seems always to be of the greatest importance I shall begin with that. I think of going from here to have a look at the Pyrenees, & if I do so may probably return to Lyons again afterward. The trip may occupy 20 or 25 days according to the number of stages I make. As I dont' know anything about the country I must just work my way on from one point to the next. The country I have been in is quite off the general English tourist beat, & I have scarcely seen any english speaking people since leaving Paris. Most of the Tourists are away in Switzerland &c. under the impression that it is or should be hot elsewhere, but the heat has not arrived yet to any great extent. I had a note from Mr Gibb in which he very vaguely suggested Russia. We had talked about it in Montreal, but I have no idea of going there. He is travelling on to travel with a Prof Budd, & though his object is a very useful one I dont see that I have any call in that direction, & as I speak neither Russian nor German I could be of very little use to the party. – I went this morning to Furviere which is a sort of suburb of Lyons, & climbed the church tower in accordance with the direction of the guide book, to obtain a view. It was very dull, however, as it happened & the view was not much. In the afternoon I went to a museum chiefly of eastern gods, from India, China & Japan. A very extensive collection, & interesting enough in its way.

I am glad you find Miss B.G. so interesting, dont think me sarcastic if I say that Mother seems to have a similar feeling. It seems to be the general impression that I should settle & marry & so on. I shall not marry now, & any kind assistance comes quite too late. One does not love twice & you cannot burn ashes back again into Coal by any known process. If I were ten or fifteen years younger I might begin life over again, but I believe I am thirty four next month, & know my own mind. I never thought life was worth much. It has taken me nearly three years to find that mine is worth nothing to me. Looked at from the best point of view, it certainly relieves one of some embarassment to have nothing to hope for, very little to fear, & no belief in anything, besides the automatic & mechanical part of me is intact, & when I get back to work as I suppose I shall some day I can go on with a certain painful exactness in the groove I have found. It is obvious one must do something to gain ones daily bread as long as one is so unfortunate as to need it, & as one grows older one becomes more & more contented to be a machine as one becomes less capable of feeling either pleasure or pain. I read somewhere lately that there are three stages in a career, first when people say he will do great things, second, he might, & third he might have. I had an up hill fight, & fought it out inch by inch till I found myself at the bottom. I never found anything worth doing that could be done without putting your whole life into it, but when failure comes under those conditions it is final. I would very much rather be happy than otherwise, but you will remember a text which says what does it matter to a man if he gains the whole world & looses his own soul.[239] Now I hope you wont mind my having written like this to you, & would rather you did not discuss the matter at all when you write. It is best, however, that you should understand, & not be at the trouble of endeavouring to gather up spilt milk. It will be a pleasure for me to do anything I can for you or the others at home.

Love to all at Metis. I hope you are having fine seasonable weather there.

Your affectionate brother

[239]Mark 8:36.

ANNA: AN ENDURING FRIEND

Seldom does one see as devoted a relationship as existed between George and Anna. He not only loved her in a very rare way but admired her many abilities and fine character. He wrote her often, offering her trips or gifts of money to help with her large family. Whenever time permitted, he told her of things of interest in which he might be engaged.

George Dawson to Anna Harrington, Ottawa, Ontario, 9 December 1881.

My Dear Anna,

When in Montreal I forgot to say that in referring to the other "volume" I left with you to read, you had made a slight mistake. The "volume" is well enough all through but the only *chapter I* specially wished you to read was that on the origen & development of Music, the subject on which we were speaking. Ottawa continues to get worse & worse & I really dont know whether I can stand it here all winter, but then, unfortunately I dont know any place I would like to go better, which makes it difficult. I hope to get into my own rooms in a day or two now which will be more settled but even less lively. There is a certain sort of satisfaction in walking about in the hotel corridors & watching the trunks come out & in. I wish I had a rapacious appetite for money or something that would keep me going. It always strikes me as particularly pleasant to be connected directly with the production of something useful say potatoes or horse-shoes & to know that if you stopped exerting yourself one day, the next you would have nothing to eat. When you write to me or send anything by post if you address. *Dept. of Interior. Geol. Survey. Sussex St.* it is unnecessary to put on any stamp as letters thus addressed to the departments go free.

Mail Closing

Yours truly

George Dawson to Anna Harrington, Ottawa, Ontario, 26 September 1893.

My dear Anna,

When I left Montreal yesterday, Mother was thinking of paying a visit to Miss Papineau at Montebello, from which I tried to dissuade her, for I think that in her present state of health it would be very risky. I do not know whether she has reconsidered the matter, but if you think as I do you will endeavour to persuade her to remain at home. On arriving here yesterday I happened to meet Mr Peter Redpath[240] at the station & got him to take lunch with me at the club. Today he was to be in Montreal & I have no doubt you will meet him. He is looking remarkably well.

I have been proposing, in an indefinite kind of way to go to the exhibition at Chicago[241] before it closes, at the end of next month, but do not feel very keen about it. If I should arrange to go could you manage to get away long enough to go also as my guest? If so it might help me to make up my mind. I have no doubt there is much to see & by taking the seeing of it not too much *Au grand serioux*, it might not be very fatiguing. I think a week at Chicago should be enough. Think of it & let me know what you decide within a day or two. I believe it would be a great advantage to you to get away from family cares for a few days – just as a change.

Yours

George Dawson to Anna Harrington, Ottawa, Ontario, 6 June 1894.

My dear Anna,

There is nothing particular that I know of to Cause me to change my general plan of leaving for the west, & I think I should be able to go on Monday or Tuesday the 18th or 19th of June. If you should decide to go out west, it would of course be very desirable that we should be able to go together.

In comparing the west & the Adarondacks I feel at a loss to express a definite opinion, though I certainly think that if a complete change of air surroundings *etc.* is desired the west offers this to the greatest extent. I do not think that the question of expense should be allowed to influence you at all, for, as I have already said I am quite pleased to be able to charge myself with that & lead the expedition if the west is decided on. The sole question should be what may be best in all probability for Eric. If Colorado is considered best, rather than Banff or some such place, the same arguments apply. The expense would not be greater, but I am not familiar with places or conditions there. These, however, are easily ascertained through doctors in Montreal who have been in the habit of sending patients there, & I feel sure that you would find no difficulty in establishing yourself there.

I have not been able to learn anything definite about Banff, beyond the fact that both the C.P.R. hotel[242] & Dr Brett's place are both now in operation & that the charges at the latter are about one-half those at the former. It would be easy to go there & find out exactly the facts. Arrangements can be made by the week or longer period at both of these places at reduced rates. But in any case the question of a few dollars more or less should not & need not stand in the way of securing the best Conditions. I think too that you should economise your own strength as much as possible, for you must have so many things to consider & provide for in arranging for the needs of the family under the circumstances. I enclose a cheque which I hope you will employ for this purpose & for any other preliminary expenses. You will require to take or send it the bank in which you may have deposit & write your name as endorsation on the back. There is no hurry about your letting me know your exact plans beyond the limit of time stated, but if you go west I should like to know a day or two in advance in order to see about tickets *etc.*

Yours

[240]Peter Redpath (1821-1894) was a prominent Montreal merchant and philanthropist whose family owned a large sugar refinery. Redpath was a major contributor to McGill and his gifts included a museum and a library. After retiring in 1879, he and his wife went to live in England at Chislehurst Manor House. They had no children that survived them.

[241]The Chicago Exhibition of 1893 was held in celebration of the four hundredth anniversary of Columbus's coming to the Americas. The exhibition, covering some 580 acres in Jackson Park on the shores of Lake Michigan, attracted a huge number of foreign exhibitors and was visited by over twenty-one million people.

[242]The Banff Springs Hotel which had opened its doors in the spring of 1888.

George Dawson to Anna Harrington, Ottawa, Ontario, 8 October 1894.

My dear Anna,

I got back here on Saturday evening, having to return unusually early in view of Dr Selwyn's visit to England. Dr S. got away in fact before I got back, but not long. Rankine I happened to meet at Revelstoke & we travelled as far as Ottawa together. He stayed here yesterday & this P.m. went on to Montreal, proposing a visit to you & Eric later. He seems to be in fairly good spirits & appears to like travelling in a professional capacity better than settled existence. & to think that he can get more travelling opportunities if he courts them. If he should select this mode of life there is really nothing against it & one does not feel inclined to influence him to give it up for a 'settled' existence which might not prove to be Satisfactory. When you finally decide what is the best course & place for Eric in the winter I trust that you will let me know what I can do to help you in the matter, & as I have already said that the question of expenses will not be allowed to interfere with whatever appears to be the best course.

Before long I shall try to make out a visit to Montreal, but for a few days it will be necessary to remain here & get things in train again

Yours

George Dawson to Anna Harrington, Ottawa, Ontario, 19 October 1897.

My dear Anna,

Many thanks for your long & interesting letter received the other day. I had a talk with William yesterday by telephone & was glad to hear that when he left Montreal you were on foot again. Can it not be arranged for you to get away for a few days still before the good weather ends, for a change & rest? This is one of the things that I should be so extremely glad to provide for if I only Knew how to do it. I can get quarters for you here if you would like to pay a visit to Ottawa & bring one of the children with you & it would be very pleasant to have you here for as long a time as you could spare. Let me know if you can manage this or any other scheme of holiday & it need not cost you anything.

I quite agree with you that the children should have all the educational chances possible, & here again, if there is anything that I can provide for I shall be only too happy to do so, but you must let me know what is required for I am not in a position to determine this. You spoke of singing lessons for Clare[243] when I was in Montreal. Now if you think that these would be useful let us begin with that & just let me know what the course will cost, including everything. Do not hesitate to do so, for it will be a pleasure to me. There is no particular pleasure in saving money till it is too late to be of any use to anyone.

At the moment I am particularly anxious that you should have a breathing time yourself if circumstances permit, so think this out without delay & submit a plan, – if not a visit to Ottawa then to some other place. Please also insist, if you can, that mother should secure all necessary help. It is most unfortunate that Father should have suffered this last attack just before the winter & when he seemed to be so much better, but it will never do to have Mother run down as well by waiting on him unassisted.

Yours

George Dawson to Anna Harrington, Ottawa, Ontario, 3 December 1899.

My dear Anna,

I have your very kind letter, with enclosures which I return to you herewith, you should keep them as they are such genuine tributes to father's memory from those who have been more or less influenced

[243]Clare Harrington (1880-1967) was one of Anna's children.

by him. Enclosed is also a note I have written to Nurse Bullock, which if you consider it suitable, you might kindly send to her.

I have also received your copy of John, & have looked into it, but am not sufficiently familiar, I fear, with the old version to clearly apprehend the benefits of the new. I will return it to you shortly. The Revelation has always a certain fascination about it, because of its diction & figurative quality, but the new rendering is not any more comprehensible to me than the old. Where does that passage about a stone taken out of a great mountain come in – I have been unable to find it.[244]

The circumstances & Mother's state of mind & health must decide the matter of her going away for a time & I hope to consult you about this please let me know what you think.

It is a great comfort to know that you have been near at hand & able to help so much during all father's failing years, & even long ago you were able to work with him & help him in a way none of the rest of us could. It has been a sacrifice to you in various ways no doubt, but one which I feel you have no cause to regret.

I hope soon to see you again.

Yours

George Dawson to Anna Harrington, Ottawa, Ontario, 14 August 1900.

My dear Anna,

I have pretty well decided to leave here for the west on Monday or Tuesday next, the 20th or 21st. Would it be possible for me to induce you to make the trip with me as my guest? It would give you a chance to see something of the mountains *etc* & would be little more than a run out & back as my business will not necessarily take more than a few days & will not require me to go as far as the coast. It seems to be a good chance & would be a great pleasure to me.

I have a letter from Rankine written yesterday in which he tells me he has postponed leaving Metis till Wednesday or Thursday. This will not give us very much time to look over MS of biography,[245] as I can only give scraps of time to it here. You might let me know what you think about it – I feel still strongly impressed with the probable utility of passing it through the hands of some outside literary man. It is likely to be the only extended biography & should be made as good as can be while we are at it. I dont think a little delay more or less matters very much.

I hope you will consider my plan for a trip favourably & take Mother's advice on it.

Ask Mother if she is sure Mr Brown (mentioned in autobiography) came from Worcester. I want to find his initials but cannot find him named in the B.A.A.S[246] lists for 1883 or 1884

Yours

George Dawson to Anna Harrington, Ottawa, Ontario, 21 August 1900.

My dear Anna,

I regret that you cannot accept my invitation for the West, although I fully realize your difficulties & agree that you are probably right in your decision. I should have been very pleased if you could have come & seen something of the Country there that I feel as much interested in as if it all belonged to me. Perhaps another chance will occur.

[244]Actually, the passage is Daniel 2:45.

[245]The reference is to a biography of their father eventually published as the rather insipid volume J.W. Dawson, *Fifty Years of Work in Canada: Scientific and Educational*, ed. Rankine Dawson (London: Ballantyne, Hanson & Co., 1901).

[246]British Association for the Advancement of Science.

About the biography I have just written to Mother & she will no doubt tell you what I have said. Rankine is to send the corrected version, typed, to Metis, & I hope that both you & Mother may now be able to go over it all & record & state your views. I think it wants only a little more to make it an interesting book now, but from the expert hand of some unimpassioned critic or revisor.

I do not favourably consider asking the now Duke of Argyll[247] to write an introduction. He has not made a sufficient mark either in literature or science to make this appear appropriate, except as a sort of advertising scheme. He might write something very good, but it would be sneered at by the first reviewer, because of the duke's position. Remember I esteem & like him very much, but he is not a natural leader & is badly handicapped by his exalted position.

The principal large changes made in father's original have been the cutting down of the 'higher education for women' to about one-half & the leaving out of a number of pages of biography of James McGill.[248] The chief thing wanting seems to be a resumé of the work father actually did, apart from educational work, represented by *Acadian Geology*[249] & his many original researches on scientific Subjects.

Please tell Ruth[250] that I have asked Mr W.J. Topley[251] here to send her the enlargements of her photo. of Poppy to Metis as soon as he can, & to return the negative to her with them. I have been unable to get them from him so far, he is so very busy.

I hope to leave tomorrow & shall be away about a couple of weeks.

Yours

George Dawson to Anna Harrington, Ottawa, Ontario, 12 December 1900.

My dear Anna,

Many thanks for your kind attention to my requirements in the way of guidance for Christmas. There is the chair for Geo.,[252] something for Mother I really do not know now what it should be since the entre dish has failed. These may each cost abt $10, so I am enclosing a cheque for $30 the balence to cover something for the younger children under Clare's guidance. Of course I shall be glad to go further if anything really desirable should turn up. I will bring down small sums of money for the older children.

Thanks also for hint abt. Eva. – I have almost given up sending her Christmas things, they require so much foresight, but I will see if I can find something.

I have had a rather disturbed week so far. Was asked by [phone] to go to dinner at Govt. House the same evening on Sunday Pm. Had my overcoat stolen out of the Club hall on Monday evening. Most of yesterday looking after Whymper the alpine man[253] who came here

247John Douglas Sutherland Campbell (1845-1914), the ninth duke of Argyll, had been governor-general of Canada from 1878 to 1883.

248James McGill (1744-1813) was a prominent Montreal fur trader and businessman who was also deeply involved in public service. His bequest of money and his estate lands for the establishment of a college were the foundations for the establishment of McGill University.

249J.W. Dawson's important geological work, *Acadian Geology: An Account of the Geological Structure and Mineral Resources of Nova Scotia...* (Edinburgh: Oliver and Boyd, 1855).

250Ruth (Harrington) Fetherstonhaugh (1882-1913) was Anna's fourth child.

251An Ottawa photographer.

252George Eric Harrington, more commonly called Eric, was Anna's oldest son.

253Edward Whymper (1840-1911) was a well-travelled, well-known British alpinist and author.

wanting some information. Evening at Naturalists Club[254] Meeting to accept a portrait of Billings[255] which is being presented to the Survey. This morning at the Police court, the Coat & the thief having been found etc etc.

Yours

George Dawson to Anna Harrington, Ottawa, Ontario, 16 December 1900.

My dear Anna,

I yesterday received a very extraordinary letter from R., which I am sending to Mother, with the suggestion that she should show it to you. The way he looks at things makes me almost doubt his sanity. All I have insisted on is that publication should be *postponed* until we are satisfied that the best has been done & have had a chance to do it. There is absolutely no reason for publication within a month or a year <if> & every argument for suitable delay if it can be bettered by it. Besides, the thing was really left in my hands & although I was very glad to have R's collaboration & got on very well with him as long as he acted in a gentlemanly way, I never resigned or intended to resign some control & oversight.

After thinking over it, I have today sent a cable to R. as follows, as a last chance. –- "Cable before Thursday agreement to postponement publication until considered here, preventing legal injunction otherwise necessary."

If he will not consent to wait, the only way I can stop the thing now seems to be as above & I believe it can be done in that way.

I hope my Christmas commissions will not bother Clare too much.

Yours

[254]The Ottawa Field-Naturalists' Club was formed in March 1879 to promote systematic study of the natural life of the Ottawa region.

[255]Elkanah Billings (1820-1876) was palaeontologist of the Geological Survey of Canada from 1856, when he was appointed by William Logan. Billings' interpretations of fossils had contributed much to the growing success of the survey.

Waterfall, date uncertain. watercolour.

Cape Ray, Newfoundland, 1869. wash.

Silver Islet, Lake Superior, May 13, 1873. watercolour.

Looking Down the Fraser River from Quesnel, B.C., 1875 or 1876. watercolour.

Miles Cone & Gordon Island, Northern Vancouver Island, June 3, 1878. wash.

Mount Gill & Whale Passage from Wright Sound, Northern B.C. Coast, September 7, 1878. watercolour.

Port Essington, Skeena River, B.C., June 5, 1879. wash.

View in Mountains, 10 Miles North of Kootenay Pass, B.C./Alberta Border, date uncertain. watercolour.

A LOYAL SON AND SIBLING

The closeness of George to his family, throughout his life, is clearly revealed in his letters to various family members. Although unafraid to challenge his brothers forthrightly and critically, George constantly remained loyal and ready to assist them whenever asked. Also, he was especially concerned about his mother's emotional and physical well-being, often admonishing her to take care of herself. The following letters illustrate George's depth of affection for and willingness to help other Dawson family members.

George Dawson to Margaret Dawson, Montreal, Quebec, 1 July 1880.

My dear Mother,

I feel I should answer your kind letter of the 24th, & though there is really nothing to say must write a note to send down by Bernard who leaves tomorrow. Everything goes along smoothly enough here & time slips away so fast that I was quite surprised to find that it is {(or rather was yesterday for it is now early morning of the 2nd)} today William is to leave you at Metis. Today (or rather yesterday) was one of these plagued holidays which make life miserable by breaking into ones routine without providing any pleasant change of occupation. I saw a St Jean Baptiste procession in town, which I suppose should count for something. I cant say I find being here in the hot weather at all equal to being in the woods, not that I feel the heat, but the solitude is more oppressive because less natural. The people one meets are no more to you than so many nine-pins. Father has has been having a dig at his fossils, which affords him great satisfaction. The foundations of the new museum are too now actually being excavated.[256] I did not see Eva's letter received the other day but from what Father says fear that she is (or was) likely to go straight to London in the first instance. This the Crowes may feel shabby, & she may also have planned to pay her Liverpool visit last, with the purpose of meeting me there. If so it will almost be better for me to put off leaving till some time in August, go straight to the Brit. Assn., then to the continent, & lastly pay any Scotch visit which may be necessary. It is, however, impossible to plan just now. I will try to visit Metis *en route*, but my stay there must be very short.

My shirts have come back from wherever you sent them to be repaired. How can I find out what is to pay on them?

With love from your

affectionate son

[256]The Redpath Museum which is located on the McGill campus.

George Dawson to Margaret Dawson, Ottawa, Ontario, 24 May 1891.

My dear Mother,

Last Sunday I was in Montreal & at the time spoke seriously of writing to you, but the idea did not materialize.

Today is again Sunday & I write but merely for the purpose of marking time & indicating that you might expect to hear from me if there happened to be anything say, as there is really nothing to communicate. I was glad to know that you had a pleasant passage & did not suffer very much on the way, but noticed by the telegrams that you must have experienced most unseasonable weather after getting to England. Here we have had all sorts of weather, cold one day & boiling hot the next & now it is becoming very dry & the air has been more or less smoky for a couple of weeks in consequence of forest fires. Today I have been taking a complete holiday, by way of a change & returned late from a long twenty-four Mile drive into the Country with Nellis, having at the outer end of the drive romped all over a hill on which pits had been dug for 'phosphate'. Generally speaking I endeavour to ballast a holiday by paying a visit of a few hours to my office – not particularly for the purpose of doing anything there but merely with the object of preserving a basis of continuity & habit of continuing to live. I shall have to go down to Montreal on Wednesday for the meeting of the R.S.C.,[257] of which meeting you have no doubt heard, & stay there possibly three days – perhaps the last visit before starting for the west, which should occur not later than the middle of next month in order that I may do anything there. My last year's report has not appeared above the horizon yet, but I have long ago given up all hope of finishing it in decent time & as no one will miss it except myself I have decided that it is not worth worrying over. It makes me feel an awful fraud when people present me with LL.D's etc & others congratulate me on same while what I do is only done because I feel too stupid to engage in anything else & take too little interest in the whole affair to get out of the rut.

My love to Eva.

Yours

[257]Royal Society of Canada.

George Dawson to Margaret Dawson, Ottawa, Ontario, 29 September 1893.

My dear Mother,

I have had a couple of notes from Father within the week in which he speaks of your improvement in health, but not in quite such definite terms as I would like. I trust, however, that you are really getting stronger & that you will take every care not to overexert yourself in any way. You were very wise I feel sure to abandon the Montebello trip, under the circumstances, though I fully sympathise with your disappointment in having to do so, for the country is looking very pretty now & it is quite the best season to make a visit. I think that you should not neglect to take a little claret every day as I feel sure that it is just such a stimulus that you require to aid digestion & keep up your general strength.

Some days ago I wrote to Anna suggesting that if she could make up her mind to go to Chicago in October I might be induced to do so also – she of course going as my guest. I have not yet had a reply <by> but suppose the matter requires some consideration.

Yours

George Dawson to Margaret Dawson, Ottawa, Ontario, 24 October 1893.

My dear Mother,

Eva's & R's letters returned herewith. I fear it is rather late to reconsider the Chicago visit now, as the exhibition is I believe to close on the 31st, next week. It seemed to be impossible for Anna to get away at the time & under the circumstances, but perhaps some plan may be devised of getting her off somewhere during the winter for a few days, for I am sure that a little change would be of great service to her.

I have no idea what R. means to convey, but fancy that he has merely been in a depressed state when he wrote. It is very easy to be depressed when you have nothing to do but think of your own grievances, real or imaginary, but it is very foolish to choose such times for writing to your friends. I shall write to him some day this week, but will not of course refer to you or his letter to you. I dont' think that you should take the matter to seriously. You can only give him good advice of a general kind & hope that he may act upon it. He has a way of looking at the dark side of everything & seems to take a perverse pleasure in doing so

Yours

George Dawson to Margaret Dawson, Ottawa, Ontario, 8 November 1893.

My dear Mother,

I return enclosed R's letter received from you this morning. It is the same old story, but if he is utterly averse to going on with medical work, or as he puts it, unable to do so, he had better much drop it & recognise the situation. If he has any small certain income from his directorship in the B.C. mine he would require to cut down his expenditure so as to live on it for the present & look out for something more or some other employment to add to it. If he thinks this cannot be done, then he had better give it up & return to Canada, where at least we can offer him means of subsistence – I should be glad to do so myself, on a moderate scale. Here he can surely find some employment, clerical or otherwise which does not need special knowledge & by which he can make an honest living. His idea of staying in London & existing penuriously there seems to me absurd, for while he is very economical in some things he is extragavant in others & is a born speculator. If he had $5000 today the chance is he would at once launch out into some "perfectly safe" thing which would leave him the same amount in debt after a few weeks. I must confess that I am quite hopeless about him. If he had strength of mind to cut all his club & other connections & drive a cab it would be much more promising, but he seems to prefer to endeavour to make an appearance on nothing & naturally finds it very hard to do so. Supplying him with money seems to be like pouring water into a sieve, but if it must be done it would seem to be better to send him a monthly allowance or a quarterly, payable at stated intervals. In this I should be very glad to join & can afford to do so. As long as he remains in his present frame of mind, he is as he says himself, one of a numerous class, who, when their friends find they will not do anything at home, ship them off to the furthest place they can think of, usually New Zealand or South Africa, sending a remittance from time to time & always hoping thay they may eventually get into something in a new country. Some of them do, when they get down to bed rock & hard manual labour. If R. feels as he writes that to practise his medical profession would be more or less of a fraud I should never urge him to do so, but at times he speaks & writes hopefully of it & if he gives it up he might afterwards make it a new cause of regret. It is quite certain that his present lonely mode of life is a most unwholesome one & leads to his exaggerating all his troubles. It is not right or fair that you should have to bear the responsibility for him, nor just that you should supply him with money from your own resources. It may not be advisable to show his private letters to Father, but I think you should at least consult with him as to R. If I can be of any help to you, I shall be very glad & quite willing to contribute, but I do not wish to encourage R. to contract liabilities which he is or may be unable to meet. Please write & let me know what you think.

Yours

George Dawson to J.W. Dawson, Ottawa, Ontario, 12 June 1894.

My dear Father,

Yours of the 10th received today. I suppose we must follow medical advice & I can see obvious difficulties in the matter of the west for Eric, but confess to being prejudiced in favour of it on general principles. I cannot say that I at all agree with your idea of remaining in Montreal during the summer for any purpose. Rankine's movements are much too indefinite to be taken into consideration & it would never do for either you or mother to remain in Montreal during the hot weather which seems now to be setting in. Metis may be a little austere & raw yet, but I feel that it will be an advantage to you to migrate there as soon as may be convenient & perhaps to look forward to returning to Montreal somewhat earlier than usual in the autumn. I suppose that mother does not seriously contemplate a journey to Banff this summer, but if she could entertain the idea I should be very glad to take charge of her & see her safely out there. I shall probably start some day early next week though I have not yet decided the exact time.

Yesterday I sent you two of my saving bank books & trust that you may have received them safely. There is no hurry about returning them to me, as they will be quite as safe in your hands as in mine.

I have received my Montreal Savings bank book back, properly made up today.

Yours

George Dawson to Rankine Dawson, Ottawa, Ontario, 19 December 1894.

Dear Rankine,

I wrote to you a few days ago addressing Bournmouth & as I note by yours of the 10th that you have returned to London it is doubtful whether you will receive my last. I note what you say about Chinese & Japanese.[258] The news from that quarter is by far the most important at present. Till a few months ago G.B., the U.S. & Russia were the three candidates or aspirants – without knowing it – for the world-dominion which history shows is sure to come. Just so soon as Japan began to break up China the U.S. were ruled out, for the time being. They were unable to do the one thing necessary to carry the day i.e. to secure the ruins of China & organize them (much as India) giving an Asiatic preponderance.

Now comes this perfectly fatuous *raprochement* with Russia by England, into which Roseberry[259] or the P of Wales or both, as for newspapers, have been hoodwinked, by which England – even by fighting perhaps – will hand over the premiership of Asia to Russia. Look at the situation. – At first Japan was & seemed to Russia to be, pulling chestnuts out of the fire for her. Russia needs Port [Sazaref] & most of Manchuria for her trans-Siberian line of railway & post. She professed to step in & take what she wanted in due time, but the continued successes of the Japs made Russia feel that she could not, so far from head-quarters, face the Japs herself successfully. The Japs were proving too strong. They were not only pulverizing China but propos-

[258]A reference to the Sino-Japanese War of 1894-1895 fought over predominance in Korea. Japan won a decisive victory, proving to be a major military power.

[259]Archibald Philip Primrose, 5th earl of Roseberry (1847-1919) was a statesman and Liberal politician, then British prime minister. After the Sino-Japanese War ended in 1895 Russia, France, Germany and Great Britain indeed scrambled for concessions from China and to keep Japan in check. The Russians received Port Arthur and the Liaotung Peninsula while the British secured a lease of the Kowloon territory on the mainland opposite Hong Kong.

ing to *keep* the gained territory. Only one thing remained to be done i.e. get England in line to help to pull these chestnuts out of the fire, get England to help to stop the Japs & when the day of settlement arrives obtain for Russia the *pied a terre* on the Pacific – after that is safely Secured & the trans-Pacific Ry. is *built* then dispose of England in the East. Thus we see Russian diplomacy falling into Lord Roseberry's arms, with tears, very much to his surprise & pleasure & the prospect of a combined intervention against Japan, acting in the interest of Russia. It seems to me, at a distance, that England should let Russia alone, make terms directly with the Japs & an understanding that they (the Japs) should have a free hand to some given line but that if they established a protectorate or annexed the territory to that line, England would step in & assure a protectorate of the rest of China. France would seize something of course, adjacent to Tonkin while Russia would have to get what she could under the circumstances from the Japs, whom she dare not incite to hostility. Germany & Co. are nowhere unless England foolishly cares to 'recognise' & thus create their interest.

Yours very truly

George Dawson to Margaret Dawson, Ottawa, Ontario, 9 February 1896.

My dear Mother,

Thanks for your kind letter of Friday. W & Florence seem pretty nearly to have concluded their 'exhaustive' search for a house before consulting me materially about it. I advised W to consult Nellis & his partner Monk & to let them make any arrangements for him & draw up proper legal agreements *etc.* if decided on. I have not seen the house they have in view, but it is entirely new & has not even got the walls tinted or papered, which alone is enough to make one shiver. It is about one mile from W's office, in a fairly good part of the town which is supposed to be improving in value. I gather that there is little or no ground about it & I dont' see how W is to go home for lunch unless he takes the street cars every day. However he promised to give the owner an answer on Saturday & has very likely by this time concuded his bargain. I took him & Florence to see my [...] idea of a house – an old low-built square stone substantial house with a veranda all round it & a piece of ground with some trees. You will see what he says on reflection, in his note enclosed. The gas work is all rubbish, for the house is on the edge of a hill, with a good view & the gas-work is down in a flat part of the town – out of sight. But I fear Florence is determined on a 'modern' house & when I see it I expect to see it topped with wooden frills & painted bed-posts. It is naturally open to suspicion for it is built to sell by a builder who makes that sort of thing his business.

I have told W that I shall lend him what money he wants, with pleasure, so that there need be no bother about that & Father need not undertake to provide any of it.

W. says he is feeling much better in health. Has been taking some sort of tonic which Wright prescribed for him. I have advised him to change it after a while for your claret again, of which I gather he has not yet consumed nearly half. He insisted on taking "Glo's"[260] letter which you sent me to show to Florence & promised to return it to you, which I hope he has done.

Yours

George Dawson to Margaret Dawson, Ottawa, Ontario, 7 March 1899.

My dear Mother,

I hope that everything is going on fairly well with you since I left

[260]Gloranna, Rankine's wife.

last week. I have been pretty busy ever since, but otherwise have nothing particular to report & have not seen W or Florence, although I reported conditions on my return.

I have been quite convinced, on thinking over it, that you should endeavour to get a few days of change & rest now somewhere & there should surely be some way of arranging for this. Could you not get Clare, for instance, to go with you for a short time somewhere, where you could read & take short constitutionals, getting up late & going to bed early. Experience leads me to believe that plenty of sleep is one of the best cures for everything. Do think of some plan of this kind & if possible let me help to carry it out. If you care to come to Ottawa for a few days I should be very glad & would be happy to secure you good quarters. Let me know if I can do so.

The club[261] is giving a dinner to the Governor General on Thursday & I have got a bidding to dine at Govt. House on Saturday, so that the remainder of this week is pretty well broken up by entertainments for me.

I have been thinking also of Father's unfortunate proclivity for writing – It is difficult to know how to treat it, but if possible I think it would be best if it can be done to firmly veto the production of the articles rather than to intercept them when produced. Father now certainly needs leading rather than to be followed & humored & I fancy he might take kindly to it. If you can give me any clue I shall be glad to write advising him.

Yours

George Dawson to Margaret Dawson, Ottawa, Ontario, 31 May 1899.

My dear Mother,

Many thanks for your note of yesterday. – I am glad to know how matters are going at home, now that you are back, also to be informed as to your plans for Metis & I do hope that you will soon safely & comfortably reach that resting place. Father derived so much benefit from the sea air there last year, that I trust he may be looking forward to his return with pleasant anticipations. Please tell him that I duly returned the list of papers to the Royal Society, with a few needed corrections.

I will do my best to visit you in Montreal on Saturday, before your flitting, though very busy with one thing or another. I have to go to Toronto on Friday of next week if I possibly can, to show a due appreciation of the honorary L.L. D. degree which the university there proposes to confer on me. It is really unlooked for & not a little embarrasing – I do not know who started or carried out the nomination but I hope that you, or someone, may take some pleasure or comfort out of it.

Yours

[261]The Rideau Club where Dawson was a prominent member and spent much time when in Ottawa.

CONTRIBUTIONS TO QUEEN AND COUNTRY: THE BERING SEA COMMISSION

Not the least of George Dawson's public service was his work on the Bering Sea Arbitration. The arbitration was undertaken to settle the limits of where the United States and Great Britain could hunt seals in the Bering Sea. This serious dispute even drew in the Russians who claimed exclusive jurisdiction over the Bering Sea by right of discovery. Their claim, however, was overruled by the argument that Spain had no exclusive right to the Atlantic Ocean even though Columbus had first reached the Americas.

Dawson was one of the commissioners sent by the British government in 1892 to the North Pacific Ocean, on an extended cruise to inquire into the conditions of fur-seals. Subsequently, he took part in conferences held in Washington, and assisted in preparation of the British case which was laid before the Bering Sea Arbitration Commission at Paris. His evidence and forcible arguments undoubtedly secured for the British case a much more favourable finding than would otherwise have been obtained. Lord Alverstone, lord chief justice of England, said: "It is not possible to overrate the services which Dr. Dawson rendered us. I consulted him through out on many questions of difficulty and never found his judgment to fail, and he was one of the most unselfish and charming characters I have ever met. I consider it a great pleasure to have known him." In recognition of his services on the arbitration, George Dawson was made a Companion of the Order of St. Michael and St. George (C.M.G.), being decorated by Queen Victoria herself.

The following are some of George Dawson's letters written while he was active on the Bering Sea Commision.

George Dawson to Anna Harrington, Unalaska Bay, Alaska, 23 August 1891.

My dear Anna,

On arrival here day before yesterday I got yours of July 27th from Metis, for which I am much indebted – also other letters of which one from Father, dated 28 July is the latest. All the news seems good, so far as it goes. I think I wrote to Father from the Pribylov Islands,[262] whence we had a pretty direct chance of sending a mail *via* San Francisco, so that you should have news of our movements by this time. Since leaving the Pribylov Islands we have made a rather extended cruise to the Northward, calling at Nunivak Island, St. Matthew Island, St Lawrence Island & eventually at Plover Bay on the Siberian coast. Tomorrow we hope to leave here for a cruise to the westward. We should be back here some time in September – proba-

[262]Pribilof Islands.

bly about the middle of the month – & hope to reach Victoria again not late in October – perhaps early in the month. Weather so far has been rather fine on the whole, for though often very foggy & occasionally wet, we have had no heavy gales. All the way South from Plover Bay we had rough sea, swell rolling about in all directions, as we afterwards found, in consequence of a storm to the southward & eastward. One day, for about eight hours, we had the worst time of rolling I ever saw, everything loose or that could break loose tumbling round all over the ship & quite impossible to cook in the galley or keep anything on a table. The fog causes a good deal of loss of time & often necessitates the expenditure of from six to 24 hours poking about every time we have to make the land. Up at St Mathews Island we found three men had established themselves for the purpose of spending the winter & hunting. The island is, however, otherwise uninhabited. Polar bears are supposed to abound on it & we saw about a dozen at different times, several being killed. We also had some shooting at walrus & sea-lions along the shore, but without any known result. The enormous numbers of sea-birds on some of these isolated islands is most remarkable. Another notable thing is the entire absence of trees or even bushes of any kind. Even here along the Aleutian Islands the steep hills & mountains of which they are composed are wet, & green with a luxuriant growth of grass & herbage wherever it is not too rocky, but absolutely treeless. There are several volcanoes along the Aleutians supposed to be 'active' but two which we have sighted so far on rare occasions of partial lucidity have carried only little languishing brushes of steam about them.

There is quite a little fleet collected here at the present moment, according to arrangements made for comparing notes *etc*. It includes three British war vessels, the Nymph, Porpoise & Pheasant, four U.S. war vessels Mohijan, Thetis, Alert & Marion. One U.S. revenue cutter & several other vessels with coal *etc*. The Pheasant has been cruising most of time in company with the Danube, & the Porpoise is to follow us when we leave here.

The above may serve as a sort of outline of news & though I address it to you, you will doubtless communicate it to the rest of the family circle as if I wrote more I should only have about the same items to reiterate.

I find that I have only this Geol. Survey paper with me, which will account for my using it instead of more suitable vehicle

Yours

George Dawson to Anna Harrington, Washington, D.C., 17 February 1892.

My dear Anna,

I have just received yours of yesterday's date & think that your apologies about writing should have been made by me, for I have scarcely begun to do my duty in that respect since I got here. Yesterday I had in view the idea of writing to wish you many happy returns of your birthday, but something intervened, & I can only hope that it may not yet be too late. I do hope that you have got that ideal nurse at last, but can scarcely hope with any assurance of being right. I shall be glad now when we can get away from Washington, for it becomes a little tiresome. So far I have had time for nothing but work of one kind or another & none to see people I want to see. Business of various kinds with returning some calls to people who have left cards here, seems to take up all the day. It is interesting for a time to be en rapport with the inner movements of diplomatic & similar matters, but it becomes tiresome. Naturally I have avoided all engagements I can, having so much to do, but Sir George[263] is a bad companion in this respect, for he seems never to be quite happy unless involved in a whirl of some kind. You will easily understand that I am only too happy to let him do all the social part of the business, but I cannot

[263]Sir George Smyth Baden-Powell (1847-1898) was a prominent British politician and author who accompanied Dawson on the 1891 Bering Sea cruise. The British claims were founded on their reports. In December 1892, Baden-Powell was appointed British member of the Joint Commission in Washington and in spring 1893 he advised in the preparation and conduct of the British case before the arbitrators in Paris.

quite escape. Mr. Hubbard,[264] for instance, who is I believe father-in law to the Bell telephone & apparently a very wealthy person, very kindly gave a dinner to the Behring Sea Commrs. from which I have just returned. He had taken pains to secure a number of well known people & we are doubtless very much indebted to him. Enclosed is a diagram of the party, which may amuse you. I dont' yet know when I shall be able to leave, but have in view the idea of visiting Montreal on way to Ottawa.

Yours

George Dawson to Anna Harrington, London, England, 19 May 1892.

My Dear Anna

Many thanks for your kind letter, received some days ago. My chief work here is in connection with report & does not so much depend upon seeing people as on pegging away in quiet, other matters counting more or less as interruptions. Nevertheless I have seen some people & made a few calls, with more I am sorry to say in prospect I called on the Marquess of Lorne[265] last Saturday & received a very kind invitation to lunch, on Sunday with him & Princess Louise,[266] I being the only guest. Last night I dined at Lord Derby's[267] – a small party, but very stately. Ld. Derby is you Know Ld. Stanley's[268] brother & is supposed to be possibly the richest man in England. Tonight I am to dine with Phillips, my old School of Mines friend, & in the after-noon I have to go to the Royal Society Meeting to be introduced there I have also called at the Tupper's[269] & have seen Sir Charles once or twice. All these things take time, as you may imagine, but now fortunately I think the most of the heavy work is over & while waiting for a few days I shall perhaps be able to [...] calls etc. I have a provisional invitation from the Marquess of Lorne to spend two or three days in the Hebrides, where he is going next week & may accept it if there should be a sufficient interlude. If I can arrange this I hope to be able to make a call *en route* at Edinburgh, where Ella Kemp is very anxious that I should come. Several people here have enquired very kindly about Rankine, as to his whereabouts *etc*, but so far I have not happened to hear of anything which would be to his advantage, or in his line

Yours

George Dawson to Anna Harrington, London, England, 10 June 1892.

My dear Anna,

Your letter of May 30 reached me yesterday, & a note from Ruth this morning. I have not yet heard anything from Rankine, though I suppose he should be arriving about this time – neither you nor Mother mentioned the steamer by which he sailed. I shall try to do what I can for him when he gets here, & hope to find that he may be inclined to buckle too at Something, even if it is only hospital practice

[264]Gardiner Greene Hubbard (1822-1897) was a lawyer and associate of Alexander G. Bell, important as business head of the early telephone companies.

[265]John Douglas Sutherland Campbell, who became the ninth duke of Argyll in 1900.

[266]Louise Caroline Alberta (1848-1939) was the sixth child of Queen Victoria who had married the marquess of Lorne in 1871.

[267]Edward Henry Stanley, 15th earl of Derby (1826-1893), was a renowned British statesman and politician who was the first secretary of state for India and later British foreign secretary. He also managed his huge Lancashire estates with great skill.

[268]Frederick Arthur Stanley (1841-1908) had been made a peer as Baron Stanley of Preston in 1886 and succeeded his brother as 16th earl of Derby on the latter's death. He was at this time governor-general of Canada.

[269]Sir Charles Tupper (1821-1915) was a prominent Canadian politician, diplomat and prime minister. At the time he was high commissioner to Great Britain.

for the time being. I have had to do a good deal of loafing of one kind or other this week, being without settled employment & waiting now chiefly for Mr Tupper's[270] arrival. Last night I went to 'Venice in London' & was quite surprised at the grand state in which this exhibition is got up. It quite takes the shine out of the real Venice in everything but actual size. One section of the show consists of a series of canals & bridges with houses along them partly built & partly painted, & narrow *dry* streets lined with stalls & shops. The canals are all connected & filled with water which is not only real but looks clean & blue. Another & the main feature is an enormous amphitheatre, with a stage at one side at least as long as the whole front of the college buildings. Between the seats & the stage is a large basin of water, which connects with the 'canals' & into which as part of the performance fleets of gondolas & barges of the most gorgeous character enter. The performances are not like ordinary plays but spectacular effects assisted by music & dancing &c &c, which succeed each other in rapid succession, & the whole stage with all the floating gondolas etc are thronged with performers, counted by hundreds.

I have promised to go tomorrow, Saturday, afternoon to Chislehurst to see Mr & Mrs Redpath & will probably return on Sunday afternoon. The weather is fine & today for the first time has been really a little warm. I hope very soon to be able to sail for Canada & am really only waiting now for a while to make sure that I am not longer needed

Yours

George Dawson to Anna Harrington, London, England, 27 July 1892.

My dear Anna,

I do not remember where I left off in records when I last wrote home – I consider all members of the family as a corporate body. Last night I had the honour of attending a magnificent banquet which Sir Donald Smith[271] gave to the Canadian Bisley rifle team at the *Hotel Metropole*. It was quite in princely style, & a number of people I Know turned up. On Monday, last week, I dined at the Redpath's at Chislehurst, a rather large dinner. On the following day I dined with Mr Ayer at the Savoy Hotel. This was a sort of *hors d'ouvres*. Nellis, from Ottawa is travelling with Ayer, who comes from New York & is a several times millionaire, the son of the author of the frills etc. This was a most lavish & gorgeous entertainment, got up entirely regardless of expense & from which I got away with difficulty about 1 A.m. It was just a little amusing too in some ways, though it is not fair to criticise one's host. – I find that I have transposed the days of the two above entertainments. I think too that in reading what I have written you will jump at the conclusion that I have been indulging in a vast amount of gaiety, but you must remember that this accounts for some ten days or so, & that most of my time has been devoted to work which has been steady & not very interesting. I have seen Rankine every few days, & hope that he is getting on satisfactorally, though he seems to find it hard to apply himself especially to an object. We went together to the Cooperative the other day & R. purchased the suits

[270]Sir Charles Hibbert Tupper (1855-1927), Sir Charles Tupper's second son, was a distinguished lawyer and statesman, then Canadian minister of marine and fisheries, who also strongly argued the British case.

[271]Donald Alexander Smith (1820-1914) was a fur trader and diplomat, best known for his financial support of the Canadian Pacific Railway.

you wrote about. They arrived later & are now in my keeping. The overcoats I did not think I had packing room for, & besides the kind of coats which would be suitable for Montreal I should suppose it difficult to get here. I hope to leave soon for home

Yours

George Dawson to Anna Harrington, London, England, 10 December 1892.

My Dear Anna,

Here I am back in my old quarters, with the bells of St Margarets, Westminster, ringing out the usual prolonged Saturday evening chime, & the Clock tower of the houses of parliament noting the hours & quarters just as though I had never left the place last August. This completes another of circles one seems to make, returning on itself & cutting out a great loop of intermediate experiences. Leaving New York early last Saturday morning, I reached London at 3.30 this (Saturday) afternoon, which makes a pretty good record.

Hope[272] turned up at the station at Liverpool this morning in time to have a short chat, but I could not wait there, as it seemed necessary to get in to London with all due speed. I shall hope to spend Christmas at Rock Ferry[273] & then to deliver the various parcels with which I am loaded.

In New York, Mother undertook a general repacking of my hand bag & distributed the gifts between it & my carry-all *etc* in such a way that even had it seemed desirable to unload these upon Hope, it would have been difficult to do so. Curiously enough, we found all the country about New York Covered with snow, & though the weather on the Atlantic was warm & pleasant, we struck a very chilly air near Ireland, a hard frost at Liverpool & a lot of snow everywhere on the inland country on the way from Liverpool to London. It is I believe snowing here tonight, but I have not been out to see

Yours

George Dawson to Anna Harrington, Paris, France, 28 March 1893.

My dear Anna,

I have now been here about a week, but have been pretty busy all the time, & have in consequence not been free from the hotel. Weather is beautiful, with tulips in full bloom in the gardens & the trees along the boulevards *etc* beginning to show fervently green. The spring is rather too early, however, & it is probable that there may still be a relapse into cold weather. Before we came over here, rooms were arranged for in this hotel, & we make up quite a large party. Mr. Tupper, as British Agent for the Arbitration, Sir C. Russell,[274] Sir R. Webster,[275] Mr. C. Robinson (of Toronto)[276] & Mr Box, as counsel. C. Russell (Junior)[277] Messrs Fronde & Maxwell[278] (secretaries), Mr

[272]Hope T. Atkin, Eva's husband.

[273]With Hope and Eva, at their home near Birkenhead.

[274]Sir Charles Russell (1832-1900), then attorney-general of England, argued the British case with great vigour and skill. He was later in 1894 appointed lord chief justice of England.

[275]Sir Richard Everard Webster (1842-1915) later became lord chief justice upon Russell's death, but was not as highly regarded intellectually as his predecessor.

[276]Christopher Robinson (1828-1905), a Toronto lawyer who presented an excellent summary of the British case, became one of Canada's most prominent lawyers.

[277]Charles Russell (1863-1928), Lord Russell's son and solicitor to the British agent in the arbitration, later became a distinguished litigator.

[278]Richard Ponsonby Maxwell (1853-1928) was a Cambridge graduate who had entered the British foreign office in 1877, where he later rose to the position of senior clerk.

Pope,[279] sec. to Mr Tupper, & myself. While to add to the number Mrs Tupper, Mrs Pope, two miss Websters, Lady & Miss Caron are also here. Macoun[280] has got quarters for himself in some less expensive establishment, but spends most of his days here.

I had a letter from Mother the other day in which she tells me that it has been definitely decided to sell "the property" to the Medical Faculty, & this I fear may be a serious inconvenience to you & Bernard. I hope that you may be able to arrange for the change in some convenient way & that it will not take place till some time after the session is over. I wrote to you some time ago about Father's probable decision. From what I hear, he seems to think that he must continue to do the work as Principal till it may suit the governors to make new arrangements. I am not quite sure that this is wise, for I can scarcely conceive of his adhering strictly to such work, & fear that it may really come to mean that he returns to endeavour to resume his old work. This I think should be avoided. Again, when it becomes known that the arrangement is only *ad interim* he would be pestered by Johnson,[281] & all other aspirants.

I write this in a room full of talk & noise, & in the intervals of conversation, & so cannot vouch for its continuity or coherency.

Yours

George Dawson to Anna Harrington, Paris, France, 27 May 1893.

My dear Anna,

I have not I fear written to you for some time but try to write to some one at home every week or so, even if there is nothing of particular interest to say. It now begins to look as if the arbitration proceedings might come to an end sometime, but scarcely I fear before about the middle of June. Sir Charles Russell is still speaking, as you are not doubt daily informed by the newspapers, but he will probably close his argument in one or at most two days more. Sir Richard Webster will then probably occupy a few days & Mr Robinson is to follow, but very briefly. The U.S. then has a right of reply by Mr Phelps[282] & whether he will be short or long I do not know, but this should end the arguments unless the court expresses a wish for some further matter – which I should think they would be very chary of doing. It is very interesting to see how all the points one has heard talked over actually come out in the spoken argument, & there are of course many other interesting items connected with the business – rumours & suggestions heard from all kinds of sources & points suddenly turning up which require reference to London or to Canada. I have not been at all continuously busy of late, but have never been quite free or in search of other occupation & so cannot be said to have particularly increased my knowledge of Paris, though we might very pleasantly spend a week or two here in simply loafing or sightseeing & amusements.

Last Sunday I lunched with a Mrs Crawford, who is the Paris correspondent of the Daily News & met a lot of rather interesting people there – Interesting as specimens of various kinds, <&> partly French partly belonging to the Anglo-French community of a miscellaneous kind which is somewhat large in Paris. There was some very excellent singing by two young ladies, whose names I forget, who are in training for proffessional musicians. Also one day this week, a party of us went to the Palais de Justice to see the Antheropometric Method in use there for the purpose of identifying prisoners. All the persons convicted of any crime are carefully measured & described by an

[279]Sir Joseph Pope (1854-1926), who had entered the Canadian civil service in 1878, was later Canadian under-secretary of state for external affairs from 1904 to 1925.

[280]John Macoun (1831-1920) was assistant director and naturalist to the Geological Survey of Canada, having assumed those positions in 1887.

[281]Alexander Johnson was at various times a McGill mathematics professor, vice-principal, and dean of the Faculty of Arts.

[282]Edward John Phelps (1822-1900) was a Yale law professor, United States ambassador to Great Britain from 1885 to 1889, and American counsel in the arbitration.

elaborate but easy system, & photographed in full face & profile to a given scale. All the particulars are entered on a printed card in each case, to which the photographs are attached, & the thousands upon thousands of resulting cards are filed away in drawers very like an immense library catalogue, in such a way that beginning with the length of the head, & following with the breadth etc you gradually eliminate. The two measurements referred to probably get rid of three quarters of the whole number of cards, & with the new measurements in hand it does not take more than five minutes to discover whether the individual in question has ever been convicted before & to connect him with all his previous record. We had for demonstration a man charged with a murder, who did not deny a previous conviction for robbery, but if he had denied it, it would not have done him any good, for he was measured & identified without any delay or difficulty.

Yours

A WIDE-RANGING MIND

George Dawson wrote with ease and clarity, and his range of subject matter was remarkable. During his working life he composed more than fifty reports as well as articles for newspapers and scientific journals. Many of his reports are technical and, therefore, not of interest to the average reader but others, on a variety of topics, could not fail to attract the interest of a thoughtful person. It will be appropriate as well as informative, then, to set down a few paragraphs from Dawson's general interest writings.

"Note on the Occurrence of Jade in British Columbia, and its Employment by the Natives," *Canadian Record of Science* 2 (1887), 365, 368-69.

My attention has been specially drawn to the use of jade by the Indians, by the occurrence of two partly worked small boulders of that material on the lower part of the Fraser (at Lytton and Yale respectively), and the discovery in 1877, in old Indian graves near Lytton, of evidence that the manufacture of adzes had there been actually carried on. These facts seem to point to the valley of the lower Fraser or to that of its tributary, the Thompson, as one, at least, of the localities from which jade has been derived, though, so far as I am aware, it has not yet been found *in situ* in any part of British Columbia. The partly worked boulders to which allusion has been made, are more particularly described below. They resemble in shape and size the well rounded stones which are abundant in rough beaches along the more rapid parts of the Fraser River, and present a peculiarity in polish which is often found to characterize these stones, and which appears to be due to the action of the sand which is drifted by the wind along these beaches during periods of low water. All the circumstances, in fact, tend to show that they may have been picked up on the immediately adjacent banks of the river.

...The peculiar adaptibility of jade to the manufacture of implements is shown by the mode of working it which has been in use by the natives, which is clearly indicated by specimens from different parts of the whole region from the Fraser River to the Arctic Sea. A suitable fragment having been discovered, it has evidently been carefully sawn up into pieces of the required shape and size, the sawing having been effected either by means of a thong or a thin piece of wood, in conjunction with sharp sand. This rude method of dividing the stone must have been very laborious, and produced a widely gaping cut before any great depth was obtained. From the fragment of a boulder obtained at Lytton (Fig. 2) [in article] flat pieces intended for adzes have been sawn off, the cuts having been carried in from the surface, on each side, till it became possible at length to break the central rib by which the piece to be detached was still united to the main mass. The boulder from Yale (Fig. 1) shows the same process in an earlier stage, though deep cuts have been made on both sides of the stone,

one of which is shown in the illustration. Several of the adzes or chisels show that the same method of sawing was adopted to trim off the edges of the flat pieces first obtained, and to render them parallel sided. Pieces thus cut from the edges of adzes are represented among specimens from graves near Lytton. Figure 3 [in article] represents a selvage piece of this kind, which has been sawn through on two sides. Figure 4, [in article] presents front and side views of a small adze from the same place, the edge still showing the median rib between two opposite saw-cuts, which has not been ground of.

Having been thus roughly blocked out by sawing, the surfaces of the adze were next generally ground flat. In the more finely worked specimens, this subsequent grinding has almost or altogether obliterated the original shaping furrows, and the surfaces have eventually been well polished.

"Note on the Distribution of Some of the More Important Trees of British Columbia," Geological Survey of Canada, *Report of Progress 1879-80*, 168B.

The flora of British Columbia as a whole may be broadly divided into four groups, indicating as many varieties of climate, which be named as follows: the *West Coast*, the *Western Interior*, the *Canadian*, and the *Arctic*. The first, with an equable cilmate and heavy rainfall, is characterized by a correspondent luxuriance of vegetation, and especially of forest growth. This region is that west of the Coast Range, and is well marked by the peculiarity of its plants. In a few spots only–and these depending on the dryness of several of the summer months owing to local circumstances–does a scanty representation of the drought-loving flora of the Califoranian coast occur. The second is that of the southern part of the interior plateau of the province, and presents as its most striking feature a tendency to resemble in its flora the interior basin of Utah and Nevada to the south and the drier plains east of the Rocky Mountains. It may be said to extend northward to about the 51st parallel, while isolated patches of a somewhat similar flora occur on warm hill-sides and the northern banks of rivers to beyond the Blackwater. In the northern part of the interior of the province, just such an assemblage of plants is found as may be seen in many parts of eastern Canada, though mingled with unfamiliar stragglers. This flora appears to run completely across the continent north of the great plains, and characterizes a region with moderately heavy rainfall, summers not excessively warm, and cold winters. The arctic or alpine flora is that of the higher summits of the Coast, Selkirk, Rocky and other mountain ranges, where snow lies late in the summer. Here plants lurk which deploy on the low grounds only on the shores of Hudson Bay, the Icy Sea and Behring's Strait.

"Mineral Wealth of British Columbia," *Journal of the Royal Colonial Institute* No. 5 (April 1893), 339-40.

Circumstances have favoured the development of the mines of the Western States of the Union, but it is, as nearly as may be, certain, that the northern half of the similar region will eventually prove equal in richness to the southern, and that when the mines of these Western States may have passed their zenith of productiveness, those of the north will be still increasing in this respect. The explorations of the Geological Survey of Canada have already resulted in placing on record the occurrence of rich ores of gold and silver in various places scattered along the entire length of the Cordilleran region in Canada, and though so far we have to chronicle only an awakening of interest in the southern part of British Columbia, these discoveries stand as indications and incentives to further enterprise to the north.

While the remote and impracticable character of much of this northern country places certain obstacles in the way of its development, on the other hand the local abundance of timber and waterpower in it afford facilities unknown in the south, which will be of importance whenever mining operations have actually been set on foot.

No attempt has been made in this brief sketch of the mineral wealth of British Columbia to enumerate the various ores and minerals which have so far been found within the limits of the province in any systematic manner. Nothing has been said of the large deposits of iron, from some of which a certain amount of ore has already been produced, and which wait to realise their true importance, merely the circumstances which would render their working on a large scale remunerative. Copper ores have also been discovered in many places. Mercury, in the form of cinnabar, promises to be of value in the near future, and iron pyrites, plumbago, mica, asbestos, and other useful minerals are also known to occur. In late years, platinum has been obtained in alluvial mines in British Columbia in such considerable quantity as to exceed the product of this metal from any other part of North America.

While, therefore, the more important products of this western mountain region of Canada are, and seem likely to be, gold, silver, and coal; its known minerals are already so varied, that, as it becomes more fully explored, it seems probable that few minerals or ores of value will be found to be altogether wanting.

Respecting the immediate future of mining, which is the point to which attention is particularly called at the present time, it may be stated that coal-mining rests already on a substantial basis of continued and increasing prosperity; while the work now actually in progress, particularly in the southern part of the province, appears to indicate that, following the large output of placer gold, and exceeding this in amount and in permanence, will be the development of silver mines, with lead and copper as accessory products. The development of these mining industries will undoubtedly be followed by that of auriferous quartz reefs, in various parts of the province, while all these mining enterprises must react upon and stimulate agriculture and trade in their various branches.

Because a mountainous country, and till of late a very remote one, the development of the resources of British Columbia has heretofore been slow, but the preliminary difficulties having been overcome, it is now, there is every reason to believe, on the verge of an era of prosperity and expansion of which it is yet difficult to foresee the amount or the end.

SELECTED POEMS OF GEORGE MERCER DAWSON

A Flower Beside the sea.

A growing flower, beside the sea;
A year, beside eternity
A wish, beside the ruling laws
Of all things governors & cause
A life mid universal death
One planned for, wished for, hoped for breath.

Thus do we fight against an iron wall
And throw our destanies on the rock {(in the teeth)?} of fate
That answers only by the death of all.
And even as we die, our hearts still long,
As wildy for the things that cannot be.
As if stern fate would yield, & right the wrong.

Grow on sweet flower beside the
    murmuring sea
And throw thy tinted petals to the <...> {sun}
Would that I were, or could be like to thee.
Grow on, beside the universal tomb
Nor heed if the expectant billows roar
Ere all things have the deep for their
    vast grave
Thou wilt be gone, nor ever wilt be
    more

Man feels that he in all things is a slave;
To time, to every law –
Laws are his dungeons, & the chains they have
The laws that curb the mind.

G.M.D.
August 1870 –

Indian Summer Reverie

How the autumn winds are gleaning
Gleaning gently in the trees
Here a leaf, & here another
Floateth down along the breeze

Blue the autumn haze is lying
In the vallies round the hills
Tinted leaves are fickly flying
On the placid woodland rills.
Fallen leaves are ever gathering,

Running circles in the vales,
Filled with their mysterious talking
Whispered songs & whispered tales

Songs & tales of by-past Summer;
How they basked at restfull noon;
Of the stars whose silence speaketh,
Of the dew beneath the moon.

But my mind is roaming, roaming,
Now along the forest glades,
Now, within the peopled city
Or in palace's arcades

Musing, musing, musing ever
On the passing sands of time,
Sometimes sunlight sometimes shadow,
On the virtue, on the crime

On the ever restless moving
On the hurry, too & fro;
On the hating & the loving
That is ceasless here below

Sometime singing sometime <laughing> {weeping}
Now in laughter, then in tears
Still the sands are sifting, slipping. –
Turns the glass, & marks the years

G.M.D.
Sept 1870 –

She is waking, She is waking
Where the distant waters flow;
Where the silent is day breaking?
& the heavens are aglow

Where the breezes that have slumbered
Shake their drowsy wings again
& arise with scents unnumbered
From the flowers where they have lain.

It is noonday here & toiling
And the floods of life are strong
And the dusty highways teeming
With the busy moving throng.

But the west is in its morning
And the shadows still retreat
While the dawn in its adorning
Follows fast with shining feet.

Feb. 72.
GMD.

Skeena River June. 1879

Down through the defiles of the hills
To seek the western ocean shore
Swift in the moonlight glancing on
Or dark in Cañons, with a roar
That in the æons does not fare.
No petty torrent pouring out
The waters of a single vale
But masterful & great thy flow
& broad & deep is writ the trace

by thee, of time upon the face
Of this wide land.
Yet to no peopled city's gate
Dost thou hear <of a > {on} the merchant freight
Nor by broad field & fertile mead
Where patient lowing cattle feed
Hast thou thy way.
A thousand nameless streams
that spring
By shattered crags & snow-fields bare
That high in alpine valleys sing
Or onward dark in forest fare
Uniting, Kissing, one by one
With <water> {current} dark or <Current> {water} clear
Through broader valleys still thy come,
By lodge of beaver haunt of deer
By Indian camps & scattered huts
When (in) thy {full} stream grown {is} <broad> {rough} & wide
(No more the hills are [miss] and dam)?
Till, happy rest! thy waters touch
The pulse of ocean & are still
Or move but with a {that} gentle throb
As the world waters sink <&> {or} fill

G.M.D.

A grove of tall & silent pines
Where moss receives the tread,
Or where the shadow darker lies
Are piled the leaves of seasons dead.
A summer sun, a seeming calm,
But to a quicker ear the roar
Of jostling atoms as they crowd
At every leaflets open pure
How soon we cease to miss the news
The noisy chatter of the day
Of battles won & lost, of games
That knaves & dupes devise & play
Then on the leafage of the time
The transient doers of today
That fill the armies of the dead
& year by year are swept away
& as they come, & pass with noise,
The peace of god continues here
& flux of time is muted out
In wooden cycles, year by year.
Peace River. Aug. 1879.

N.W. Coast V.I. 1885.

A dream of rocks with yeasty waves {seas}
& storm-wrecked trees along the shore.
Of strong sea winds which shape the clouds
In silence, while the breakers roar
My point of view, this light canoe
That dips & rises in the foam,
& seeks along each surges crest
A landing place & point of rest.

The sea & its Song. Outer Coast of Vancouver Island, 1885. –

To sleep {rest} on fragrant cedar boughs
Close {Near} bye the western ocean's view
While {When} in the tops of giant pines
The livelong night the sea-winds hymn,
And low upon the fretted shore
The waves beat out the evermore –

The hard fought margin of the deep
Where all the waving forces sleep.
Tis thus that life is full content
And still the world is [...] & wide
This night, the stars, by heaven sent
And I & whatsoever betide.
No discord breaks the perfect whole
The sea reflects but one refrain
Sings sleep, – sleep, – sleep oh weary soul,
Sleep – ask not if there wake again.

From my Tent. Fraser River, Above Lillooet
5. Oct. 1889.

A fire that twinkles on the hill,
Dim mountains rising tier on tier;
Thin mists, below, the valley fill,
And over all the full moon clear.
The slumbrous sound of cricket's song
That drones & drones the whole night long,
While deep below, with steady roar
The river frets its rocky shore,
From everlasting, evermore.
A parched balsamic-laden air
That still flows warm as in the day
Tired horses cropping scanty fare
Along the slopes of sun-baked clay —
Oh psalm of life & death & time
Whose solemn music beats so true
And fills the soul with dim unrest —
Despair of that we know & do.
Thy symphony is drawn afar
From some remote full orbed star
That from the depth of primal night
Sends {Sheds} but one {limpid} ray of light.

Just as a wee maid when she stands
with downcast eyes & folded hands
to say her oft conn'd task
So blushing on some mossy bank where days are
long & woods are dark,
or crowded thick twixt lichened stones
where some old glacier laid his bones;
Those nodding bells are swung
Fairer than all <&> {where} all are fair,
within the flowery band
& breathing out a perfume rare
where the tall ranked pine trees stand
In the lone distant northern land

Contorted beds of unknown age
<My> {Our} weary limbs shall hear,
Perchance some neat synclinal fold
At night may be <my> {our} lair.
Dips <I> {we} shall take on <unknown> {unnamed} streams
Or where the rocks strike follow
Along the crested mountain edge
Or anticlinal hollow.
Where long neglected mountains stand
Fast crumbling into shreds
And laying bare on every hand
The treasures of their beds
We'l gently with the hammer wake
The slumbering petrefaction
That for <some> {a} hundred million years
Has been debarred from action.
Or snatch some Crinoid or mollusc
Unearthed without our toiling

Adrift upon the river bed
By brute attrition spoiling
[...] one day in bring back
Into the sunlight glory
All natures misbegotten <types> {shapes}
of pattern rude & hoary
To reptile of prodigious bones
or two tailed salamander
To wed the lovely name of Jones!
{Gives} <To> Jones good cause to wander.

In common with British subjects around the world, Dawson was deeply stirred by the Boer War, in which his fellow countrymen played so conspicuous a part. His view of the war, however, was not wholly one-sided; his admiration for the Boers being evidenced by the following:

The silent boer, that lies, a clod –
He was a father or a son –
Upon his dry grey Transvaal sod
Among the rocks that we have won;
His narrow soul was true & strong,
To fend us from his home & kraal
He gave his life – we know him wrong,
But find him worthy after all.
And when in days to come the song
Of later harvests shall be sung,
He will have part in that south land
As elder brother true & strong.
Each spring that rises on the veldt
Will cost its wreath of self-sown flowers,
Will breathe its fragrence & be felt
About his grave & over ours.
Not all is lost if life be spent
For it is good to truly die
To give to that extreme extent
If so be freedom lives thereby.
The things not seen, beyond the veil,
Have harvest also full & true
And loss {gain} we reckon but by toll
Is measured there – To each his due.

U.P. Ry. or Central Pacific Ry.

A Station in the parchéd west
Between two lines of wrinkled hills
Whose seams are dry & show alone
The finger-marks of ancient rills.
The sleepy balmy scent of sage
That floods the baking desert plain
A noise of coming wheels, a stir
Of drowsy folk to see the train
A hundred windows blinking past
& then the silence falls again

Life is ever uncertain, which my uncle realized when he wrote:

Sic transit gloria mundi

Life is a bubble on the sea,
The ocean of eternity.
It floats a while in glittering pride,
It may o'er many billows ride.
There comes a moment, none knows why.
No cloud o'erspreads the summit sky;
Some little breath, some hidden thing,
Perhaps a spirit on the wing –
Touches the orb – it melts away. –

The sea receives its little spray. –
No mask, no memory, left behind;
The everlasting sea, the wind –
Flow on.
GMD. March 13th. 1870.

Life

At best a poor sleep-waking,
A consciousness of pain
In which we strive to know & do
& striving sleep again.
A sound of distant voices
That talk within the night
The twitter of awakened birds
Before the morns broad light.
We argue that a dawn must be
That Questionings must {shall} answer find,
That in some chain of being linked
Is the dim wistful human mind.
But they who do no question hold
Who live in the corporeal {are alone in outer} sense,
Must they move on to higher things
Nor sink to utter nescience?
Nay rather, happy day remains
To those <those> who live not for the night,
But true & striving faithfully
Shall win a way from light to light

Misinchinca River B.C.
July 22. 1879
G.M.D.

How those we love we pity most
We see in guise of every day
The surging upward of the soul
Within its envelope of clay
We note the path of rapid years
In growing furrows whitening hair
But find no word of full reply
To loose the gird of petty care
There still is longing unexpressed
Some latent wealth divine of love,
Some dream of an idyllic rest
Or undersigh for things above.
Which finds no voice or answer here
No image in the changing year
No concord in our little day

9 Sep. 88.

Great God, the father of mankind,
The spring of life, the hand of fate;
I bow to thee in humble mind
& kneel before thy golden gate
That bars the <most> {sun} this close of day
One star above the mountain crest,
The dark & utmost verge of earth, {all}
That drops full swift into the west
Upon the footsteps of the day.
A thousand stars that start behind
From out the ancient realm of night

The growing darkness fills the land
& stills the thousand tongues {many tongued} of day
Tis only on my knees I dare
To look afar, or scan the way

Which I must tread, to look & pray.
& when above the path I turn
To where the lights of heaven burn
My lips refuse to utter prayer.
No [plummet] dark natures deep
Through which the swift milleniums sweep
I know not, cannot understand.
But stricken silence may express
The revered awe I must (confess?)

Vernon B.C. 1890. G.M.D.

At some time towards the end of his life, Dawson wrote this poem:

The threads of life – a tangled skein
My time is short – The threads of life,
A tangled skene, I cannot sort,
But what it gain to live –
To live & die. To see & know
& pass to the unknown.
If <stand> I might live <again> {anew}, & plan
Throughout & shape again
So far as man may do
The web of life – would I
Or would I not pursue
The self same scheme?
Would I be led away as heretofore
Or rule my life anew
& weave new dreams?
I know not, for it even seemed to me
That I chose well & truly,
<a> {that} default was made, not so much
Or above by men, as by <the> an
Overruling fate.
One must be god like, or a god
To <...> {rule}
With knowledge of the future every act.
But still I cannot think that all
Must end in failure, all must be in vain
Thought is too subtle, <love is too> to intense
To die & have no place.
Love is too deep & hope too high to fail
Of <some> their fruition somewhere <or> at some time,
{Perchance} It is but to resolve to live again
To grasp the clues of love to [sweep]
Through all the realm of darkness to some life
Which <lies> {is} beyond, which must recur
Where lies fruition, where the words unsaid,
The songs running, the <dreams> immatured
dreamt dreams <of> that glow <like sunset>
<On the world, take form> to my dim
eyes, like sunset on the world – Take form.
Where all that has been wrong, <&> {or} wrongly ordered
Will be well

Friends are made & friendships broken
Lives are woven & untwined
Loving hearts without a token
Float apart & never find
On this earth another meeting.
Though they part so very {& part} lightly
With a friendly word & greeting
Scarce a tear-drop gliding brightly
Still they part, & {mayhaps} part for ever
And their eyes & hearts will never,
Never hold commune again.

G.M.D.
Sept 29 – 1870.

My love! Dear [...] loved
so long ago
You chose your path & went another
way
I was not rich nor great, & told you so
But in my love to you could never
stray.
Within me rose, I knew, some tide
of the divine
Long purpose of the world, some pulse
of that great heart
That rules. Had you been mine
It seemed we might have lived a
life apart
Have breathed some air all consecrate
& true,
Inviolate & pure; your love to me
& mine alone to you.
But that may be no more, time past
is dead.
When last your hand left mine, that
hour
We two were parted, never
waters led
That turned two drops upon the
mountain ridge
Of some great continent was greater
bar.
Our lives diverged & ever wider spaces
spread all between, & far
Far from our childhoods place
We drift <& flow> & drift, & you
To me it seemeth <lost in barren> {left in moving} sands
<while I touch but infertile rocks>
are lost. While I, touching the barren
rocks
Go onward through grey lands
To that great sea that looks <...>
<...>
The habitable world in one embrace
God grant we these may some day
meet & face to face
For there is but one love for me & one
for you
& in some flux of time this must return
as truth is true.

Tonight I turn to my old love,
As fair as light – as false as as hell.
I gave her all there is in life
With no reserve. I loved her well,
Nay fondly, threw my ardent soul
Upon the earth that she might tread
On purple there, & safe & <well> whole
Escape the mire & clay & live.
I loved her well – But she was false,
<Inbred> {Inbred} & tainted with a lie
That virtue is not truth & may
<To> In sport decree that men shall die.
Yet if she would return to me
Though <old> {poor} & grey & false & <poor> {old}
I would receive her like a queen
& throne her on a seat of gold.
She took my youth, <with> {in} her weak hand,
A plaything for an idle day

Her good pleasure, no truth, scarcely truth.
No virtue perhaps. –
Who can say?

Did she not know so well
In her own scheming way
That the other would not in the long run repay

Tell of constancy love & romance
In a glance,
Yet, strange, in the end

Does not marry a friend
But happily finds her whole heart
        Can be given
To a stranger with money but no hope
        of heaven
            G.M.D.

SCIENCE AND EXPLORATION IN CANADA

Though always aware of the practical application of his work Dawson, nonetheless, refused to compromise his ideal of free scientific inquiry. Rigorous study of a scientific subject under exacting standards was requisite for any advancement of Canadian society. In the following, then, Dawson vigorously argued his commitment to the wedding of the applied with the theoretical.

"On Some of the Larger Unexplored Regions of Canada," *Ottawa Naturalist* 4 (1890), 29, 39-40.

Fortunately, or unfortunately as we may happen to regard it, the tendency of our time is all in the direction of laying bare to inspection and open to exploitation, all parts, however remote, of this comparatively small world in which we live, and though the explorer himself may be impelled by a certain romanticism in overcoming difficulties or even dangers met with in the execution of his task, his steps are surely and closely followed by the trader, the lumberer, or the agriculturist, and not long after these comes the builder of railways with his iron road. It is, therefore, rather from the point of view of practical utility than from any other, that an appeal must be made to the public or to the government for the further extension of explorations, and my main purpose in addressing you to-night is to make such an appeal, and to show cause, if possible, for the exploration of such considerable portions of Canada as still remain almost or altogether unmapped.

...To sum up briefly, in conclusion, what has been said as to the larger unexplored areas of Canada, it may be stated that while the entire area of the Dominion is computed at 3,470,257 square miles about 954,000 square miles of the continent, alone, exclusive of the inhospitable detached Arctic portions, is for all practical purposes entirely unknown. In this estimate the area of the unexplored country is reduced to a minimum by the mode of definition employed. Probably we should be much nearer the mark in assuming it as about one million square miles, or between one-third and one fourth of the whole. Till this great aggregate of unknown territory shall have been subjected to examination, or at least till it has been broken up and traversed in many directions by exploratory and survey lines, we must all feel that it stands as a certain reproach to our want of enterprise and of a justifiable curiosity. In order, however, to properly ascertain and make known the natural resources of the great tracts lying beyond the borders of civilization, such explorations and surveys as are undertaken must be of a truly scientific character. The explorer or surveyor must possess some knowledge of geology and botany, as well as such scientific training as may enable him to make intelligent and accurate observations of any natural features or phenomena with which he may come in contact. He must not consider that his duty consists merely in the perfunctory measuring of lines and the delineation of rivers, lakes and mountains. An explorer or surveyor properly equipped for his work need never return empty handed. Should he be obliged to report that some particular district possess no economic value whatever, besides that of serving as a receiver of rain and a reservoir to feed certain river-systems, his notes should contain scientific observations on geology, botany, climatology and similar subjects which may alone be sufficient to justify the expenditure incurred.

"The Progess and Trend of Scientific Investigation in Canada," *Proceedings & Transactions of the Royal Society of Canada* 12 (1894), LII-LIII.

We find ourselves possessed in Canada of a country vast in its dimensions, but of which the population is as yet comparatively small. If, therefore, we have good reason to believe that the natural resources of our territory are in any respect commensurate with its area, we may look forward with confidence to a great future. But in order that this may be realized properly and soon, we must devote ourselves to the exploration and definition of our latent wealth, and to the solution of the problems which inevitably arise in the course of its utilization under circumstances which are often more or less entirely novel. For this purpose, we are provided at the present day with methods, appliances and an amount of accumulated knowledge not previously thought of, but which we must be prepared to enlist in our service if our purpose is to be achieved.

It is unsatisfactory to read, as we often may, the statement that Canada is possessed of "unlimited natural resources," for such a statement means little more than that we have been unable to make even a reasonably complete inventory of these resources. In order intelligently to guide the work of those endeavouring to utilize the benefits given to us by nature in the rough, and to attract population and capital for this end, it is necessary to be much more specific. It is true that great regions of Canada still remain very imperfectly or almost altogether unexplored, but we are nevertheless already in a position to form some general estimate of the importance and character of the products which the country as a whole is best capable of yielding. Thus, in respect to mineral wealth, I believe we are justified in assuming that Canada is equally rich with any known area of the earth's surface of comparable dimensions. So, in regard to products of the sea, these, relatively to our length of coast line–and this is very great–are probably at least as valuable as those of any other similar length of coast. Of arable and pastoral land, because of the rigorous climate of the northern portions of the geographical area of Canada, the extent is not commensurate with that of the country, but it is practically so great that we may be pardoned if describing it at present as "unlimited." As to the natural wealth represented by our forests, it is probably correct to state that Canada is still capable of affording more timber than any other country in the northern hemisphere; but of this, with the constant and increasing drafts upon it, we can already begin to see the end, unless some effective measures shall be taken, and that soon, towards its conservation and reproduction. We have, in fact, yet to learn to regard a forest as a special kind of farm, in which, if we do not sow, we cannot hope to reap perennially.

It is not, however, my purpose to enter into any details respecting the natural wealth of the country, but rather to point out as briefly as may be what has been done and what still remains to be accomplished by means of the various scientific organizations and associations of Canada, in aid of the utilization of these resources, in the matter of making them known to the world at large, and toward the solution of various important questions which lie before us in connection with them. Science is but another and a convenient name for organized knowledge, and as such it has entered so largely into every branch of human effort, that when, at the present time, any one attempts to pose as a "practical" in contradistinction to a scientific worker, he may be known to be a relic of the past age, in which much was done by rule of thumb and without any real knowledge of the principles involved. Neither can we safely make any division between what is sometimes called "practical" or "applied" science and science in general, for the knowledge must be gained before it can be applied, and it is scarcely yet possible to bar any avenue of research with a placard of "no thoroughfare," as an assurance that it cannot lead to any material useful end.

At the same time, there are certain directions in which investigation is very closely wedded to results of immediate and tangible value, and it is practically in such directions that the State may reasonably be expected to exercise its activity. But the line should not be too rigorously drawn, for should the investigator for a time stray into some by-path of research, because of his individual interest in his work, it is not improbable that he may return from his excursion with some unexpected discovery, which may prove to have important bearings on the problems of every-day life. Take, for example, the study of Palaeontology which, relating as it does, to extinct forms of life, might appear to be a branch of science wholly removed from any practical object, however interesting it may be to disinter and to reconstruct these remarkable forms. But we all know that this study has become an indispensable one as an aid to the classification of the rock formations and thus to the search for the useful minerals which some of these contain. This is more particularly the case perhaps in the instance of coal beds, which are usually confined in each region to some set of strata, which may be defined with precision only by the aid of the evidence afforded by fossil remains.

THE LASTING LEGACY OF GEORGE MERCER DAWSON

On Thursday, 28 February 1901, George Dawson was at work in his office in Ottawa and dined that evening at the Rideau Club. The next day he fell ill from acute bronchitis and a telegram was sent to his mother in Montreal. She came at once, arriving in Ottawa on Saturday evening, but when she reached her son's room at Victoria Chambers, it was to learn that he had died fifteen minutes earlier. He was in his fifty-second year. One can only imagine the door of Heaven swinging wide to welcome this lonely but valiant traveller. Perhaps the sweet smell of pines and salt air greeted him, while wild flowers in joyous profusion nodded their heads in welcome!

On the following Monday, 4 March 1901, the Ottawa Free Press *printed:*

> *His death was most unexpected, and casts a great gloom and sorrow upon the whole Staff and all who knew him. He was not only high in the world of science – By his demise, there is removed from the sphere of activity one of the greatest intellectual lights of the last monument which will ever be a crown to his life work – He knew his geology and the grasp he had of all problems relating to the economic and natural resources of our vast Dominion made him master of his Department and a centre of distribution of most valuable information.– He did much to establish mining upon a firm and rational and non-speculative basis.*

The headlines of other newspapers printed the words: 'CANADA'S GREAT LOSS,' while tributes came from leaders in government, scientific societies, and universities. But, the news of his passing brought perhaps the finest tributes from men of the Canadian west. For example, the following poem by Clive Phillipps-Wolley appeared in the British Columbia Mining Record[283] *as an ode, To* "Dr. George."

Grey and ghostly willow fringes, flame to crimson at
the tips,
Where a sun that has some heart in, through the
waking forest slips.
High above us, on Mount Sicker, I can hear the blue
grouse hoot;
Bird are calling, rivers glitter; buds are bursting,
grasses shoot.
On the pine stump, by our shanty, Dawson's tattered
map lies spread,
And my partner with his finger, marks the footsteps
of the dead.

[283]*British Columbia Mining Record* 8 (April 1901), ix.

"Spring!" he says, mate, time to quit it, for the barren lands and hoar,
Where the Earth's heart freezes solid and the mighty bull moose roar:
Where through silent spaces, silent, reckless bands of hardfists hold,
By this here map and the compass, their course to the northern gold,
With a laugh and a curse at the danger, while down the Arctic Slope
Are two of the best ahead of the boys, Doctor George and Hope–

Hope she has fooled us often, but we follow her Spring call yet,
And we'd risk our lives on his *say so and steer the course he set,*
Down the Dease and the lonely Liard, from Yukon to Stikine,
There's always a point to swear by, where the little doctor's been
Who made no show of his learning, but Lord! what he didn't know
Hadn't the worth of country rock; the substance of summer snow–
I guess had he chosen, may be, he'd have quit the noise and fuss
Of cities and high palavers to throw in his lot with us.
He'd crept so close to Nature, he could hear what the Big Things say,
Our Arctic Nights, and our Northern Lights, our winds and pines at play.
He loved his work and his workmates, and all as he took for wage
Was the name his brave feet traced him, on North-lands newest page –
That, and the hearts of the hardfists, though I reckon for work well done,
He who set the stars for guide lights, will keep him the place he won,
Will lead him safe through the Passes and over The Last Divide,
To the Camp of Honest Workers, of men who never lied –
And tell him the boys he worked for, say, judging as best they can,
That in lands which try manhood hardest, he was tested and Proved A Man.

The following poem was written by George's brother William Bell Dawson, at the time of his death:

'IN MEMORIAM'

"A kindly world, has lavish honours strewn;
Ability and worth they gladly own;
The genial smile they see, the cheery tone –
The inner life lies hidden, and unknown –
Ah gentle soul and pure, and must thy path
Lead through the desert sand?
In boyhood's days, affliction's heavy hand
Upon thy form, and lifelong burden laid
Enough we think – and see in early life
The vision of a happy home to cheer
Thy loneliness – Alas the dream of youth
Proves but mirage to mock thy hope –

Now bleaker drearier still the prospect lies
Yet toil thou on for duty's path lies straight

Forgetting self – Thy thought of others now
But disappointment still in sorrow's steps
Must ever walk to keep thee company
Yet, at the end, compelled to crown thy brow
And do thee homage as their conqueror!
Their work is done, thy noble soul
With quiet patience bears the last distress
While life ebbs out, with fading close of day
Ah God the great disposer of our lives
We bow adoring, silent at Thy feet
For it is Thou, whose throne on justice stands
Where righteousness and peace together meet."

There is little left to say about this amazing uncle, George Mercer Dawson, whose indomitable and vital spirit inspired all. So let his biography be closed by these hopeful words found written on a scrap of paper in his handwriting:

"LIFE'S GOOD NIGHT is GOD'S GOOD MORROW TO ETERNAL LIGHT."

FOR FURTHER READING

Barkhouse, Joyce. *George Dawson: the Little Giant* (Toronto: Natural Heritage, 1989).

Cole, Douglas, and Lockner, Bradley, eds. *The Journals of George M. Dawson: British Columbia, 1875-1878*, 2 vols. (Vancouver: UBC Press, 1989).

Cole, Douglas, and Lockner, Bradley, eds. *To The Charlottes: George Dawson's 1878 Survey of the Queen Charlotte Islands* (Vancouver: UBC Press, 1993).

Elson, J.A. "The Contributions of J.W. Dawson (father) and G.M. Dawson (son) to the Theory of Glaciation," *Geoscience Canada* 10 (1983): 213-16.

Lang, A.H. "G.M. Dawson and the Economic Development of Western Canada," *Canadian Public Administration* 24 (Summer 1971): 236-55.

Parsons, John E. *West of the 49th Parallel: Red River to the Rockies, 1872-1876* (New York: William Morrow, 1963).

Turner, Allan R. "Surveying the International Boundary: The Journal of George M. Dawson, 1873," *Saskatchewan History* 21 (Winter 1968): 1-23.

West, John J. Van. "George Mercer Dawson: An Early Canadian Anthropologist," *Anthropological Journal of Canada* 14 (1976): 8-12.

Wright, Allen A. *Prelude to Bonanza: the Discovery and Exploration of the Yukon* (Sidney, B.C.: Gray's Publishing, 1976).

Zaslow, Morris. *Reading the Rocks: the Story of the Geological Survey of Canada 1842-1972* (Toronto: Macmillan Company of Canada, 1975).

ABOUT THE AUTHOR

Lois Winslow-Spragge was the daughter of Anna Lois and Bernard J. Harrington who was the first Professor of Chemistry and Mineralogy at McGill University.

She grew up near the McGill Campus where her grandfather, Sir William Dawson, was principal for almost forty years (1855-1893).

She inherited the family love of nature and followed the artistic talents of her mother and her uncle George, both of whose water colour paintings and sketches are now a part of the McCord Museum collections.

Lois was the first teacher of drawing at Miss Edgar's and Miss Cramp's School for Girls in Montreal, an active member of the Canadian Handicrafts Guild and one of the first female groups of Montreal potters.

Her love of rocks and minerals came naturally to her and in the 1970s she exhibited her Rock Art Paintings at the International Geologists Congress held at McGill University. At this same exhibition George Mercer Dawson's paintings were mounted and exhibited for the first time.

ABOUT THE EDITOR

Bradley Lockner is an historical editor and professional librarian with a special interest in British Columbia history. He has collaborated on two other volumes on George Dawson, and is currently working on another volume of Dawson's later journals. He has also been involved in several other historical editing projects, including the journal of the fur trader Alexander Walker, and the letters of Gilbert Malcolm Sproat, Indian Land commissioner in B.C. during the 1870s.

VISUAL CREDITS

Cover	"Sioux Camp, Turtle Mountain…,"	Anne V. Byers
	"Oxen Train leaving the Depot…,"	Anne V. Byers
	"Bridge over Souris River…,"	Anne V. Byers
	"Oxen Train and Horse Train…,"	Anne V. Byers
	"G.M. Dawson."	Anne V. Byers
Page 22	"George Dawson as a child"	McGill University Archives
Page 23	"Dawson's Parents …,"	McCord Museum, Notman Photographic Archives
Page 24	"The Dawson sons…,"	McGill University Archives
page 25	"Anna Lois Dawson, 1891…,"	McCord Museum, Notman Photographic Archives
Page 26	"At the Dawson summer home…,"	McGill University Archives
Page 91	Officers of the British Boundary …,"	Anne V. Byers
Page 92	"Scouts of the British Boundary …,"	Anne V. Byers
Page 93	"Maxim Marion Métis guide,"	Anne V. Byers
	"Trader McPherson's family…,"	Anne V. Byers
Page 94	"Ice cutting party, British Boundary …,"	Anne V. Byers
Page 95	"Red River Transportation…,"	Anne V. Byers
	"Chippewa Indians at Dufferin, …,"	Anne V. Byers
Page 96	"Camp at Turtle Mountain Depot,…,"	Anne V. Byers
	"Observation Tent Zenith Telescope,…"	Anne V. Byers
Page 97	"Sappers building Boundary Mound,"	Anne V. Byers
	"Métis Traders, 1874,"	Anne V. Byers
Page 98	"Log hut at Métis settlement…,"	Anne V. Byers
	"Depot on East Fork of Milk River…,"	Anne V. Byers
Page 99	"Dead Crow Indians on Boundary …,"	Anne V. Byers
	"Grave of Sioux Indian, 1874,"	Anne V. Byers
Page 100	"Assiniboine Indian Camp, 1874,"	Anne V. Byers
Page 112	"Skedan Indian Village…,"	PA44329, National Archives of Canada
	"Skedan Indian Village…,"	PA38153, National Archives of Canada
Page 113	"G. M. Dawson and party…,"	PA51137, National Archives of Canada
Page114	"Looking up Waterton Lake…,"	PA51149, National Archives of Canada
	"San Josef Bay…,"	PA37943, National Archives of Canada
Page 115	"Indian students, Alert Bay…,"	PA37934, National Archives of Canada
Page 116	"Boat building camp…,"	PA52787, National Archives of Canada
Page 117	"Frances Lake, Yukon,…,"	PA52717, National Archives of Canada
	"Looking up Pelly River,…,"	PA37560, National Archives of Canada
Page 118	"Kamloops, B.C. 1888,"	PA52734, National Archives of Canada
	"Hydraulic Mine, Horsefly,…,"	PA38204, National Archives of Canada
Page 161	"*Waterfall*"	McGill University Archives
Page 162	"*Cape Ray,…,*"	McGill University Archives
Page 163	"*Silver Islet,…,*"	McGill University Archives
Page 164	"*Looking Down the Fraser River,…,*"	McGill University Archives
Page 165	"*Miles Cone & Gordon Island,…,*"	McGill University Archives
Page 166	"*Mount Gill & Whale Passage,…,*"	McGill University Archives
Page 167	"*Port Essington, Skeena River,…,*"	McGill University Archives
Page 168	"*View in Mountains,…,*"	McGill University Archives

INDEX